AF307343

Technische Universität Dresden

Physics-based compact modeling and parameter extraction for InP heterojunction bipolar transistors with special emphasis on material-specific physical effects and geometry scaling

Tobias Nardmann

von der Fakultät Elektrotechnik und Informationstechnik
der Technischen Universität Dresden

zur Erlangung des akademischen Grades eines

Doktoringenieurs

(Dr.-Ing.)

genehmigte Dissertation

Vorsitzender	Prof. Dr.-Ing. habil. Mayr		
Gutachter	Prof. Dr.-Ing. habil. Schröter	Tag der Einreichung:	04.11.2016
	Prof. Dr.-Ing. Rudolph	Tag der Verteidigung:	24.04.2017

Bibliographic information published by the Deutsche Nationalbibliothek
The Deutsche Nationalbibliothek lists this publication in the Deutsche Nationalbibliografie; detailed bibliographic data are available on the Internet at http://dnb.dnb.de .

Bibliografische Information der Deutschen Nationalbibliothek
Die Deutsche Nationalbibliothek verzeichnet diese Publikation in der Deutschen Nationalbibliografie; detaillierte bibliografische Daten sind im Internet über http://dnb.dnb.de abrufbar.

© 2017 Tobias Nardmann
Herstellung und Verlag:
BoD – Books on Demand, Norderstedt
ISBN: 978-3-7448-7280-5

Alle Rechte vorbehalten. All rights reserved.
Gesetzt vom Autor.

Für meine Eltern
und für Julia.

Danksagung

Eine Arbeit wie diese hat zwar nur einen Autor, hätte aber ohne die Unterstützung vieler anderer Menschen nicht entstehen können.

Zuerst möchte ich mich natürlich bei Prof. Michael Schröter bedanken, der als Doktorvater nicht nur mit seinen Kontakten in die Industrie, sondern auch und gerade mit deinem Verständnis für die Materie stets hilfreich war und neue Anstöße zu geben verstand, wenn sie gebraucht wurden. Ohne ihn wäre diese Arbeit nicht zustande gekommen.

Herzlichen Dank sagen möchte ich an dieser Stelle auch Prof. Rudolph für die Begutachtung der Arbeit.

Dank gebührt auch allen aktuellen und ehemaligen Mitarbeitern des Lehrstuhls für elektronische Bauelemente und integrierte Schaltungen. Martin Claus, Kai-Erik Moebus und Steffen Lehmann haben mich bereits als SHK betreut und mir Werkzeuge und Arbeitsmethoden nahe gebracht, die ich immer noch nutze. Andreas Pawlak war stets bereit, die Zeit für intensive Diskussionen über HBT-Physik, Messtechnik oder Brettspiele aufzuwenden. Gemeinsam mit Gerald Wedel habe ich die ersten Versuche für die Simulation von III-V HBTs gemacht. Er hat auch Materialmodelle für mich in den Simulator CHIEF implementiert und dazu beigetragen, dass ich nicht nur im Drift-Diffusions-Weltbild denke. Mit Tommy Rosenbaum und Julia Krause habe ich mich über die Parameterextraktion austauschen können. Paulius Sakalas hat mir bei der Arbeit im Labor sehr geholfen und einige Messungen für mich durchgeführt. Mit Holger Wittkopf habe ich mir lange Zeit ein Büro geteilt und konnte mit ihm oft Fachliches und weniger Fachliches besprechen.

Frank Chen, Dr. Miguel Urteaga und Dr. Pete Zampardi möchte ich für die Transistoren danken, die sie und ihre jeweiligen Firmen zur Verfügung gestellt haben.

Neben der fachlichen Dimension haben mich viele Freunde auf dem Weg zur Disseration begleitet. Marieke Frassl hat bei der Arbeit an ihrer Promotion oft dieselben Probleme wie ich gehabt und mir mit ihrer Perspektive sehr geholfen. Andreas Herwig ist ein ausgezeichnetes Vorbild dafür, dass soziales Engagement neben der Arbeit immer möglich ist. Jessica Hoge, Adrian Mörchen, Klemens und Anja Muthmann und Markus Windisch haben durch gemeinsame kreative Hobbies dazu beigetragen, dass ich nicht nur mit Gleichungen zu tun hatte.

Meine Eltern und meine Schwester haben mich auf meinem Lebensweg natürlich nicht erst seit dem Studium begleitet ebenfalls einen großen Anteil am Gelingen dieser Arbeit. Dafür und für ihre stete Unterstützung bin ich ihnen sehr dankbar.

Ein ganz besonderer Dank gilt auch meiner Frau Julia. Sie hat stets daran geglaubt, dass ich diese Arbeit fertig stellen kann, auch wenn ich selbst nicht so sicher war und war geduldig mit mir, wenn ich zu viel über Elektronen und Diffusion geredet habe.

Abstract

Given the increasing interest in both high-frequency and high-power applications, III-V HBTs and especially devices based on InP - which can be faster than SiGe HBTs and retain a better power-handling capability - have the potential to fill an important demand in the microelectronics world. Already, such transistors have achieved maximum oscillation frequencies beyond 1 THz. A major obstacle for more widespread deployment of III-V technologies is the lack of systematic process scaling, compact modeling and model parameter extraction. III-V foundries generally offer a limited number of device sizes with individual (i.e. non-scalable) model cards for each of them. Thus, circuit designers cannot freely choose device sizes. Additionally, physics-based modeling is not a high priority, with self-heating - the main breakdown mechanism for III-V HBTs - commonly being disregarded. Finally, the formulations for dedicated III-V compact models are based on those found in the SGPM model and cannot include all necessary physical effects in a consistent way. These factors combine to prevent circuits designed in III-V technologies from reaching their full potential.

Based on an analysis of the device structure and material properties, the most important differences between III-V and SiGe HBTs are identified with the goal of developing extensions for modern compact models used for SiGe and enabling them to model also III-V devices. The HICUM compact model is chosen as a basis for this work. The impact of the NDM effect on the collector charge and the doping- and composition grading on the junction capacitance require an extension of the compact model code, which is performed in a physics-based way consistent with existing model formulation.

The parameter extraction is investigated, with a special focus on the series resistance extraction, which often employs so-called extraction methods for determining the resistance value based on terminal data from a single transistor. Those methods are evaluated using compact-model generated data with a known target value. This evaluation is based on three separate InP HBT processes for which compact models were generated. Suitable methods are identified and some improvements are suggested. An extraction example for all relevant model parameters for single-transistor extraction is shown.

An approach for scalable parameter extraction, typical for SiGe devices, but uncommon for III-V HBTs, is applied to a 250 GHz InP HBT technology. Test structures for many technology parameters have been fabricated and are evaluated in order to increase the physical accuracy of the model parameters. Problems with the scaling approach are discussed based on deviation between the assumed and the actual device dimensions, indicating a need for a more stable fabrication process.

Finally, a comparison between the compact model and measured data is shown for the scalable model based on the 250 GHz InP technology, for the fastest commercially available InP technology with $f_{\max} > 500$ GHz and for a 80 GHz GaAs technology.

Kurzfassung

Durch das stetig steigende Interesse an Schaltungen, die sowohl bei hohen Frequenzen als auch bei hohen Leistungen arbeiten, haben III-V HBTs - insbesondere mit dem Basismaterial InP, die schneller als SiGe HBTs arbeiten und dennoch eine höhere Durchbruchspannung aufweisen - das Potential, eine wichtige Lücke in der Mikroelektronik zu füllen. Solche Transistoren haben bereits Oszillationsfrequenzen oberhalb von 1 THz erreicht. Ein wichtiges Hindernis für die großflächige Anwendung von III-V-Technologien ist der Mangel an einem systematisch verwendeten Ansatz für die Skalierung von Prozessen, die Modellierung der Bauelemente und die Extraktion von Parametern für die Modelle. III-V-Hersteller bieten generell eine eingeschränkte Zahl von Transistoren an, für die individuelle - also nicht skalierbare - Modellparametersätze erstellt wurden. Schaltungsentwickler können also die Maße der zu verwendenden Transistoren nicht frei wählen. Darüber hinaus ist die physikbasierte Modellierung der Bauelemente meist keine Priorität für die Hersteller; sogar die Selbsterhitzung der Transistoren, die oft die Ursache für die Zerstörung der Bauelemente ist, wird häufig nicht mit einbezogen. Weiterhin basieren die Gleichungen der Kompaktmodelle, die spezifisch für III-V HBTs entwickelt wurden, auf dem SGPM-Modell und enthalten nicht alle relevanten physikalischen Effekte. Die Kombination dieser Faktoren verhindert, dass das Potential von III-V-Technologien voll ausgeschöpft werden kann.

Basierend auf einer Analyse der Materialien und Strukturen werden die wichtigsten Unterschiede zwischen III-V und SiGe-HBTs identifiziert. Das Ziel der Analyse ist, Erweiterungen für moderne SiGe HBT Kompaktmodelle zu entwickeln und diese damit in die Lage zu versetzen, auch III-V HBTs modellieren zu können. Das Kompaktmodell HICUM wird als Grundlage für diese Arbeit verwendet. Durch den Einfluß des NDM-Effekts auf die Kollektorladung und des Dotierungs- bzw. Materialübergangs auf die Sperrschichtkapazität werden neue Gleichungssysteme für das Modell benötigt. Die entwickelten Gleichungen sind physikbasiert und die Formulierung ist konsistent mit dem aktuellen Modell.

Die Extraktion der Modellparameter wird mit einem besonderen Fokus auf die Serienwiderstände untersucht. Letztere werden oft unter Verwendung von sog. Extraktionsmethoden durchgeführt, die messbare Klemmendaten eines einzelnen Tranistors direkt mit einem Widerstand in Verbindung brin-

gen. Diese Methoden werden auf der Basis von modellgenerierten Daten mit
einem bekannten Sollwert ausgewertet, was eine Bewertung der Genauigkeit
ermöglicht. Einige Verbesserungen für Methoden werden vorgeschlagen. Beispiele für die Extraktion aller relevanten Modellparameter werden auf der
Basis von Einzeltransistor-Extraktion gezeigt.

Ein Ansatz für die skalierbare Parameterextraktion wie er typischerweise
für SiGe HBTs, aber selten für III-V HBTs durchgeführt wird, wird für eine
250 GHz InP HBT Technologie implementiert. Teststrukturen für zahlreiche
Technologieparameter wurden hergestellt und werden ausgewertet, um den
physikalischen Bezug des Modellparametersatzes zu verbessern. Probleme
mit dem Skalierungsansatz werden auf Basis der Differenz der erwarteten
und tatsächlichen Transistordimensionen diskutiert; diese Differenz impliziert,
dass stabielere Produktionsprozesse benötigt werden.

Abschließend wird ein Vergleich zwischen Modellergebnissen und Messdaten für die 250 GHz InP Technologie, die schnellsten kommerziell verfügbaren InP Transistoren mit $f_{\mathrm{max}} > 500$ GHz und eine 80 GHz GaAs Technologie gezeigt.

Contents

1 Introduction   **1**

2 Modeling III-V based HBTs   **5**
  2.1 Introduction   5
  2.2 Differences between SiGe and III-V HBTs   6
    2.2.1 III-V material properties   6
    2.2.2 Vertical composition   8
    2.2.3 Top view and cross section   12
  2.3 Compact modeling   15
    2.3.1 Principles of compact modeling   15
    2.3.2 The HICUM compact model   17
    2.3.3 Existing III-V compact models   19
  2.4 Numerical simulations of III-V materials   21
    2.4.1 Material properties   22
    2.4.2 $n^+ - n - n^+$ structure   23
    2.4.3 HBT structure   26
    2.4.4 Conclusions   29

3 Model extensions   **31**
  3.1 Introduction   31
  3.2 Temperature dependence of the base current components . . .   31
  3.3 Collector-emitter parasitic capacitance   33
  3.4 Multi-region junction capacitance   35
    3.4.1 Standard compact model capacitance description   36
    3.4.2 Derivation for the multi-region model   38
    3.4.3 Voltage limiting   42
    3.4.4 Model verification   45
    3.4.5 Temperature dependence   47
  3.5 Collector transit time   48

  3.5.1 Low-current transit time 48

  3.5.2 Medium current transit time 62

4 Compact model parameter determination 75

 4.1 Introduction 75

 4.2 Parameter determination 76

  4.2.1 Parameter determination methodology 76

  4.2.2 On-wafer measurements 78

  4.2.3 Pulsed measurements 81

 4.3 Evaluation of extraction methods for series resistances 86

  4.3.1 Evaluating extraction methods 86

  4.3.2 Investigated process technologies 89

  4.3.3 Methods for the emitter resistance 90

  4.3.4 Methods for the collector resistance 106

  4.3.5 Methods for the base resistance 115

 4.4 Single device parameter extraction 127

  4.4.1 Junction capacitances 127

  4.4.2 Diode parameters 129

  4.4.3 Series resistances 130

  4.4.4 Avalanche current parameters 131

  4.4.5 Temperature coefficients and thermal resistance 132

  4.4.6 Transit time parameters 136

  4.4.7 Transfer current parameters 141

  4.4.8 Additional extraction steps 143

  4.4.9 Summary 144

 4.5 Geometry-scalable parameter extraction 145

  4.5.1 Technology description 147

  4.5.2 Test structures 148

  4.5.3 Geometry separation 153

  4.5.4 Scaling of other parameters 157

  4.5.5 Discussion 158

5 Model application 163

 5.1 250 GHz InP technology 163

 5.2 500 GHz InP technology 168

 5.3 GaAs technology 174

6 Summary and outlook 177

Bibliography 181

A Material parameters for numerical device simulations **197**
   A.1 Band gap 197
   A.2 Band gap narrowing 198
   A.3 Effective mass 200
   A.4 Dielectric constant 200
   A.5 Impact ionization 200
   A.6 Recombination 201
   A.7 Low-field mobility 203
   A.8 Field-dependent mobility 204
   A.9 Energy relaxation time 204

B Impact of relative permittivity on the collector junction capacitance **207**

C Supplementary figures for model verification **209**
   C.1 250 GHz InP process 210
   C.2 500 GHz InP process 216
   C.3 GaAs process 220

List of Symbols

Abbreviations

(C)MOS	(Complementary) metal oxide semiconductor
B, C, E, S	Base, collector, emitter, substrate
BE, BC, CE	Base-emitter, Base-collector, collector-emitter
BJT	Bipolar Junction Transistor
CMC	Compact Model Coalition
CNTFET	Carbon nanotube field effect transistor
DD	Drift-diffusion
EC	Equivalent circuit
FOM	Figure of merit
GaAs	Gallium arsenide
GICCR	Generalized integral charge control relation
HBT	Heterojunction Bipolar Transistors
HD	Hydrodynamic
HEMT	High electron mobility transistor
HICUM	The HICUM bipolar transistor model. High current model
HV	High voltage
IC	Integrated circuit
III-V	Material consisting of elements from the third and fifth main group of elements
InGaAs	Indium gallium arsenide
InP	Indium phosphide
MOSFET	Metal-oxide-semiconductor field effect transistor

RF	Radio frequency
SCR	Space charge region
SEM	Scanning electron microscopy
SGP	Spice Gummel Poon
Si, Ge, SiGe	Silicon, germanium, silicon germanium
SIMS	Secondary ion mass spectroscopy
TEM	Transmission electron microscopy
VA	Verilog-A

Constants

m_0	Electron mass
ε_0	Vacuum permittivity
k_B	Boltzmann constant
q	Elementary charge

Variables

β_f	Forward small-signal current gain
B_f	Forward DC current gain
C_{BE}, C_{BC}, C_{CE}	Terminal base-emitter, base-collector and collector-emitter capacitance
C_{dEi}, C_{dCi}	Internal base-emitter and base-collector diffusion capacitance
C_{jEi}, C_{jCi}	Internal base-emitter and base-collector junction capacitance
E_{jC}, E_{wC}	Electric field at the BC junction and at collector-subcollector transition
f_t, f_{max}	Cutoff frequency and maximum oscillation frequency
I_{BE}, I_{BC}	Currents flowing through the base-emitte resp. base-collector diode
I_B, I_C, I_E	Terminal base, collector and emitter current
I_T, J_T	Transfer current, transfer current density
J_B, J_C, J_E	Terminal base, collector and emitter current density
N_B, N_C, N_E	Doping concentration in base, collector and emitter
Q_{jEi}, Q_{jCi}	Internal base-emitter and base-collector junction charge

Q_{p0}, Q_p	(Zero-bias) hole charge
R_{Bi}, R_{Bx}, R_B	Internal, external and total base resistance
R_{Cx}, R_E	Collector and emitter series resistance
R_{th}	Thermal resistance
T_0, T	(Reference) temperature
$V_{B'E'}$, $V_{B'C'}$, $V_{C'E'}$	Internal base-emitter, base-collector and collector-emitter voltages
V_{BE}, V_{BC}, V_{CE}	Terminal base-emitter, base-collector and collector-emitter voltages
V_{Ci}	Internal collector voltage
V_T	Thermal voltage
w_{BC}	Collector SCR width
w_c	Collector width

CHAPTER 1

Introduction

The trend in modern electronics towards ever higher frequencies of operation and complexity as well as power efficiency requires a broad palette of different technologies to be available to circuit designers for various applications. While metal-oxide-semiconductor field effect transistors (MOSFETs) dominate the digital world, they have apparently reached their top analog performance around the 65 nm node [SUZC15, VTD$^+$13]. Emerging technologies such as carbon nanotube field effect transistors (CNTFETs) theoretically offer excellent properties such as very high linearity and speed, but have yet to deliver on those promises in practice [SCS$^+$13]; the measured transit frequency (f_T) of emerging devices rarely exceeds approximately 10 GHz [SKW$^+$11] due to the impact of series resistances, parasitical capacitances and other unwanted effects.

Heterojunction bipolar transistors (HBTs), on the other hand, offer a number of key advantages over competing technologies: a very high transconductance and therefore a relatively low impact of a load impedance on the transistor operation [SUZC15], a high transit frequency (f_T) and maximum frequency of oscillation (f_{max}) at a comparatively relaxed lithography and favorable noise characteristics.

Like all semiconductor devices, HBTs can be fabricated in different semiconductor materials. The most common are silicon-germanium (SiGe) HBTs, which even today reach values above $(f_T, f_{max}) = (505,\ 720)$ GHz ([HRB$^+$16]) and are projected to eventually reach the THz range ([SWH$^+$11]). SiGe HBTs are easily integrable with MOSFETs in so-called BiCMOS processes, which utilize the advantages of both technologies.

However, HBTs fabricated in III-V materials (i.e. semiconductors consisting of one element of the third and one of the fifth main group of elements) offer a versatile alternative. Depending on the materials that are used, III-V HBTs can be the fastest available bipolar transistors [UPR$^+$11] (competing only with high electron mobility transistors (HEMTs), also fabricated in III-V materials, for the title of fastest available transistors overall), offer very high breakdown voltages and therefore excellent power-handling capability and show good linearity [WMX$^+$06] and low noise figures at high frequencies [SS13]. The importance of a possible integration with a complementary MOS (CMOS) process in Silicon has also been recognized in the III-V community and been the subject of some research in recent years [Kaz13]. Typical applications for III-V HBTs include handset power amplifiers (PAs) (e.g. [MdRZ13]), high-efficiency [GURP15] and high-speed [RRG$^+$12] amplifiers as well as high-speed oscillators [SUH$^+$11]. Overall, III-V-based HBTs and especially Indium-Phosphide (InP) HBTs are excellent candidates for future high-speed communication circuits.

In order to fully utilize the capabilities of any transistor technology in circuit design, a good compact model is needed. A compact model allows the prediction of circuit performance ahead of fabrication, thus saving time and money by reducing design iterations. Similarly, a model that agrees very well with the actual device performance allows slimmer safety margins and therefore more optimized circuits. If the model is physics-based, that is, if its EC elements have a direct physical relation to a region of the device and its parameters can be determined from material properties, physical constants and the device geometry, it can also be used for predictive modeling, i.e. predicting the performance of future devices with some confidence. Its parameters can be determined in a relatively low frequency range, typically several GHz, while retaining accuracy up to frequency ranges where experimental characerization is difficult for small-signal and impossible for large-signal operation. Furthermore, a scalable compact model that is able to determine the device performance for nearly arbitrary sizes and configurations of transistors helps designers choose exactly the device that is needed, thus reducing power dissipation, and can even be used in circuit optimization routines.

Unfortunately, compact modeling of III-V HBTs is not on a level comparable with what is common in the silicon world. The lack of accurate transistor models is a major obstacle for deploying InP technology in production circuit design (e.g. [HZC$^+$07], [Zir12]) and is documented by various attempts to improve existing compact models (e.g. [IRS$^+$03], [NMPG09]). The most widely used compact model for III-V HBTs, the Agilent HBT model (AHBT) [Agi16], is based on the UCSD HBT model [UCS00], which is itself built on the Spice Gummel-Poon (SGPM) model [GP70] and does not contain physics-based formulations for multiple effects typically encountered in modern HBTs. Additionally, despite notable exceptions [HZC$^+$07], the use of test structures for the accurate determination of compact model parameters is uncommon. However, models intended for the SiGe material system cannot necessarily be used for III-V HBTs directly due to unique physical effects occuring in the materials as well as the vertical and lateral structure employed in their manufacture.

The goal of this work is to include important effects occurring in III-V materials in a compact model for circuit design in a physical, yet intuitive way in order to aid deployment of III-V HBTs in prototypes and products. The HICUM compact model [SC10] was chosen as a basis, since, along with MEXTRAM, it is one of two models supported by the Compact Modeling Coalition (CMC) as an industry standard and has been widely deployed in process design kits (PDKs) for years by semiconductor manufacturers around the world. The model is in constant development, striving to include new physical effects as they become important due to increasingly aggressive vertical and lateral scaling, but to date no comprehensive effort to include the effects typical for III-V materials was made. However, the models derived for physical effects in this work are applicable not only in HICUM, but for compact modeling in general. Similarly, the approaches for parameter determination for, e.g., the series resistances, the area-perimeter separation or the junction diodes are not model-specific.

This thesis is structured as follows: In chapter 2, basic principles of compact modeling and the expected differences between HBTs manufactured in the SiGe respectively III-V material systems are discussed. A summary of the changes made to the standard HICUM/L2 model is given in chapter 3. Chapter 4 focuses on the determination of model parameters from both test structures and single transistor extraction methods, providing examples for the various steps as a guide to modeling III-V HBTs, before a comparison between model and measured data for multiple technologies is shown in chapter 5. Finally, a summary of this work as well as an outlook on future work is given in chapter 6.

CHAPTER **2**

Modeling III-V based HBTs

2.1 Introduction

In this chapter, general principles of compact modeling as well as compact modeling specifically with a focus on III-V HBTs are discussed. Numerical device simulation as an important tool for understanding device physics is also briefly addressed.

In order to use some of the advantages of compact modeling, a physical basis for the models is required. This allows linking a certain measured device behavior to the appropriate parameters, enabling, e.g., scaling of the model. While many physical principles in SiGe and III-V HBTs are very similar (e.g. injection of minority carriers from the emitter into the base and subsequent transport into the collector), there are a number of differences caused by both the material itself as well as the transistor cross-section typically in use. The most relevant differences are summarized in 2.2.

Compact models for electron devices form the link between the semiconductor technology and circuit design. In modern circuits, it is impossible to accurately predict the circuit performance without the aid of computer-aided design (CAD) tools. Iterating designs over multiple wafer runs to optimize performance is prohibitively expensive and time-consuming. Simultaneously, numerical device simulations are not only too time-intensive, but also too inflexible to be used in circuit design since they rely not on extractable parameters, but a known doping profile and material parameters and serve better as a tool for understanding the semiconductor physics in principle. Physics-based compact models, which can be used in a circuit simulator, form an ideal middle ground for circuit design. They can also have additional functionality in predictive and statistical modeling, allowing yield prediction and even process debugging. The basic principles of compact modeling are discussed in section 2.3.

While device measurements are the ultimate reference for a compact model, numerical semiconductor device simulations are a very useful tool. They allow an investigation of the status within the device by solving the differential equations (depending on desired complexity and physical accuracy the drift-diffusion (DD), hydrodynamic (HD) or Boltzmann transport equation (BTE) systems for HBT simulations) describing the device behavior based on technological data such as doping profiles, known material data and biasing. If all input parameters are accurately known and all relevant physical effects are included, the solution is very accurate. For compact modeling, these device simulations are a powerful tool for understanding the devices and verifying model simplifications. Section 2.4 briefly discusses the application of numerical simulations to III-V materials.

2.2 Differences between SiGe and III-V HBTs

This section discusses the differences between SiGe and III-V HBTs. These differences indicate which additional effects may have to be taken into account in the compact model. This work assumes that the reader is familiar with the basic operation of a bipolar junction transistor (BJT) and heterojunction bipolar transistor (HBT) and does not discuss their physical background in depth. The interested reader is referred to the literature, such as [SC10, Rud06]. Instead, only major differences in III-V materials and cross-sections and their expected impact on a compact model are described.

Three categories will be investigated: The properties of III-V materials, the vertical doping and material profile and the lateral geometry. The latter primarily affects scaling equations and has an impact on the relative quantities of certain equivalent circuit elements; the former two can have a direct impact on the physical description of the device.

2.2.1 III-V material properties

2.2.1.1 Electron and hole mobilities

HBTs can be fabricated as npn-devices, where electrons as minorities in the base carry the transfer current, and as pnp-devices, in which holes are mainly responsible for the current transport. Complementary devices are especially useful in many types of digital circuits. While both types of devices can also be fabricated in III-V materials, the electron mobility is commonly very high, but hole mobility is very low, as listed in tab. 2.1. While this is also true for Silicon, the ratio of p- to n-mobility is much worse for III-V materials and pnp-devices are typically not fabricated at all.

Table 2.1: Electron and hole mobilities at low fields and dopings for various materials [Syn15]

Material	Silicon	InP	InGaAs	GaAs
μ_{n0} / $cm^2\,V^{-1}\,s^{-1}$	1417	4500	22600	8467
μ_{p0} / $cm^2\,V^{-1}\,s^{-1}$	470	150	250	356

2.2.1.2 Negative differential mobility

A unique property of III-V materials is the negative differential mobility (NDM) effect. The effect is caused by the relative energetic proximity of the Γ- to the L-valley; if electrons gain sufficient energy, they transition into the L-valley, where the carrier mobility is lower as a consequence of the different valley parabolicity. With increasing electric fields, more electrons transition into the L-valley and the overall velocity of the electrons shrinks. This is schematically demonstrated in fig. 2.1.

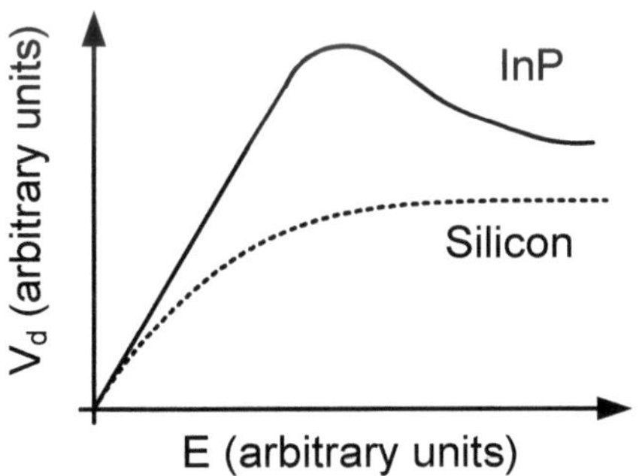

Figure 2.1: Schematic velocity-field characteristic for silicon and for InP as an example of III-V materials.

The NDM effect is relevant for carriers in medium to high fields. In a HBT, such fields occur in the collector SCR. It is expected that the collector transit time increases as the collector fields increase, i.e. with higher reverse bias at the BC junction.

The NDM effect is often confused in literature with velocity overshoot. The latter, however, has its origin in non-local transport, during which the electrons are not in equilibrium with the electric field. Carriers are limited to a certain drift velocity due to scattering. In rapidly spatially changing fields, though, a carrier can be accelerated past the velocity normally expected before a scattering event occurs. This effect does not occur in the Drift-Diffusion view of carriers, as it assumes that the carriers are always in equilibrium with the field, but is covered by HD or BTE simulations.

2.2.2 Vertical composition

Tab 2.2 shows an excerpt from the layer structure of a very fast InP HBT and is used here as an example for the discussion of differences between SiGe and III-V HBTs to be considered. The structure is, in principle, representative of III-V HBTs, showing all relevant effects. There are two things of particular note here:

- The base is made of a different semiconductor material than the emitter and the collector, i.e. both the BC- and the BE-junctions are hetero-junctions. This classifies the device as a Double-Heterojunction Bipolar Transistor (DHBT) as opposed to a Single-Heterojunction Bipolar Transistor (SHBT), which has, as the name implies, only one heterojunction. Although SHBTs have fallen out of use in recent years, the term HBT in this work can refer to either type of transistor. The purpose and impact of the heterojunctions is briefly examined in section 2.2.2.1.

- The collector is neither uniformly doped (δ doping in tab. 2.2) nor does it consist of a single material (BC grading in tab. 2.2). The impact of this is discussed preliminarily in section 2.2.2.2.

Note that SiGe HBTs, with a base consisting of silicon-germanium and a collector and emitter consisting of silicon, are DHBTs.

2.2.2.1 Heterojunctions

Heterojunctions are pn-junctions with a change of semiconductor material as well as doping type. The advantages of heterojunctions in an electron device were theoretically known before such a device could be fabricated [Kro57].

The main purpose of a Base-Emitter heterojunction is to increase the current amplification of a transistor. In a BJT, the ratio of the injected electron to the hole current - and therefore, disregarding recombination, the current amplification - is given by the doping levels in the emitter and base N_E and N_B, the carrier mobilities of electrons in the base μ_{nB} and holes in the

Table 2.2: Partial layer structure of an InP HBT [UPR$^+$11]

Description	Doping / cm^{-3}	Material	Thickness / nm
Emitter Cap	$4 \cdot 10^{19}$	InGaAs	80
Emitter	$8 \cdot 10^{17}$	InP	10
Emitter	$5 \cdot 10^{17}$	InP	40
Base	$(7 - 4) \cdot 10^{19}$	InGaAs	30
Setback	$3.5 \cdot 10^{16}$	InGaAs	15
BC grading	$3.5 \cdot 10^{16}$	InGaAs/InAlAs	24
δ doping	$3.5 \cdot 10^{18}$	InP	3
Collector	$3.5 \cdot 10^{16}$	InP	108
Subcollector	$1 \cdot 10^{19}$	InP	5
Subcollector	$2 \cdot 10^{19}$	InGaAs	6.5
Subcollector	$2 \cdot 10^{19}$	InP	300

emitter μ_{pE} (which are dependent primarily on the doping) and the emitter and base widths w_{E} and w_{B}. The current gain is calculated as

$$B_{\mathrm{f,ideal}} = \frac{N_{\mathrm{E}} w_{\mathrm{E}} \mu_{\mathrm{nB}}}{N_{\mathrm{B}} w_{\mathrm{B}} \mu_{\mathrm{pE}}}. \tag{2.2.2-1}$$

The allowable base doping for a practical transistor is limited by the desired current gain. Therefore, the base resistance is relatively large.

With the introduction of a second semiconductor material, however, differences in the conduction and valence bands are introduced. In material systems typically used for HBTs, the conduction band difference blocks the electron current, while the valence band difference blocks the hole current. In a properly chosen material system, the latter is much larger than the former; additionally, the conduction band difference only fractionally impacts the current transport due to the formation of a space charge region at the BE interface and the corresponding band bending, which leads to the energy levels W_{S} above resp. W_{N} below the conduction band. Thus, the current amplification becomes approximately

$$B_{\mathrm{f,ideal}} \approx \frac{N_{\mathrm{E}} w_{\mathrm{E}} \mu_{\mathrm{nB}}}{N_{\mathrm{B}} w_{\mathrm{B}} \mu_{\mathrm{pE}}} \cdot exp(\frac{\Delta W_{\mathrm{V}}}{kT}). \tag{2.2.2-2}$$

Typically, the valence band difference is much larger in III-V HBTs than it is in SiGe (e.g. $\Delta W_{\mathrm{V}} \approx 105\,\mathrm{meV}$ between Si and SiGe with a 15% Ge content vs. $\Delta W_{\mathrm{V}} \approx 380\,\mathrm{meV}$ between InP and InGaAs lattice matched to InP);

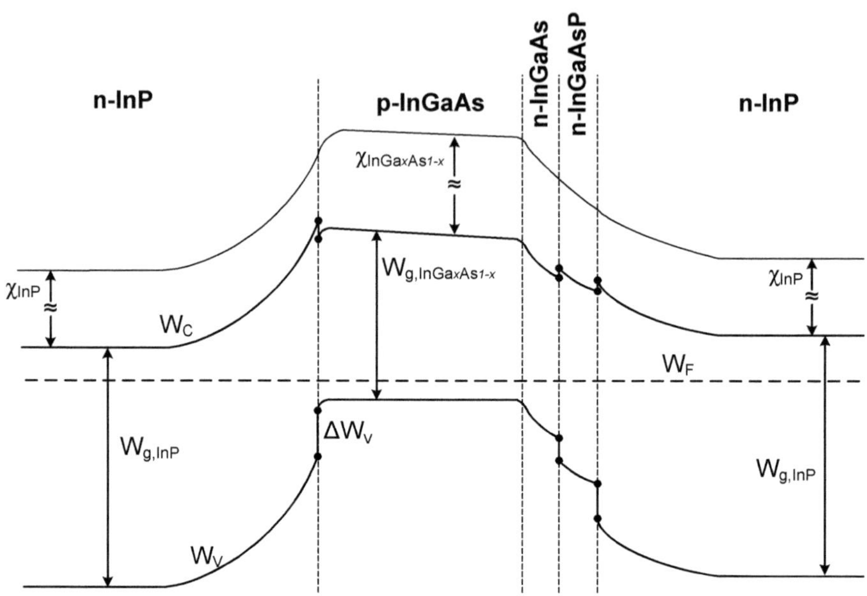

Figure 2.2: Schematic band diagram of a type-I InP DHBT at thermodynamic equilibrium with the conductance resp. valence band energies W_C reps. W_V, the Fermi energy W_F, the band-gap W_g and the electron affinity χ.

it suppresses the hole injection to the point at which volume recombination in the neutral base and surface recombination at the extrinsic base surface instead of the hole injection into the emitter are the main component of the base current [TLB07].

The spike in the conduction band, which affects the injection of electrons into the base and therefore the transfer current, results in a lower value of the saturation transfer current, effectively turning the device "on" at higher BE voltages. It also acts as a so-called launcher: Electrons that cross the spike possess a high kinetic energy when they are injected into the base, which may translate to shorter base transit times. There is, however, no agreement on this in the literature (e.g. [ZCS+04, RHFS94, DL92]). In either case, it influences only the low-bias base transit time and is therefore important for the device speed, but not compact modeling because it is already included in the low-bias base transit time τ_0.

The BC heterojunction is not implemented to exploit the material impact on the junction directly; in fact, the band-gap differences caused by the material transition are usually detrimental. Instead, the main purpose is to use a material with a high band-gap in the collector. Since an electron must have a kinetic energy at least equal to the band-gap in order to ionize a molecule and thus generate an additional electron, i.e. trigger the avalanche effect, a higher band-gap means a higher required field strength for the onset of the avalanche effect and therefore a higher breakdown voltage.

The impact of the material change on the band diagram of the HBT at the base-collector interface is, in principle, the same as for the BE junction: the formation of a barrier for either or both kinds of carriers. Unlike the BE junction, though, the BC junction is commonly reverse biased. In order to reach the collector, the electrons traversing the base must overcome the conduction band spike. The carriers that do not possess sufficient energy to do so reduce the transfer current and increase the base charge, thus increasing both the transit time and the base current through recombination. This is referred to as current blocking.

In order to avoid high spikes, the BC heterojunction is commonly fabricated as a series of material steps, splitting one large spike into multiple smaller ones, as indicated in fig. 2.2. Additionally, the transition is set back from the BC interface; the reason for this is that it allows the electrons to gain sufficient kinetic energy in the electric field of the BC SCR to overcome the spikes.

As the current rises and the electric field in the BC SCR weakens, current blocking becomes important. Initially, this effect was relevant only for III-V HBTs. In high-speed SiGe HBTs, however, the effect could also be observed and was included in HICUM/L2.

It should be noted that the above considerations are mainly valid for so-called type-I HBTs, in which $W_{C,base} < W_{C,coll/em}$ and $W_{V,base} > W_{V,coll/em}$. Type-II HBTs use a material system in which $W_{C,base} > W_{C,coll/em}$ and $W_{V,base} > W_{V,coll/em}$. Then, the current blocking effect is not relevant. The trade-off is usually a suppression of electron injection into the base, resulting in a lower transfer saturation current. Examples of the HBT types can be found in, e.g., [BFA+16, DRM+11].

2.2.2.2 Collector doping profile

A thin region of increased doping in an otherwise uniformly doped collector, listed in tab. 2.2 as δ-doping, is common for III-V HBTs. Its purpose is twofold:

- Linearization of the BC junction capacitance and therefore increasing the overall device linearity and

- increasing the current-carrying capability of the device by increasing the electric field strength at the BC junction.

Especially the latter point is important for the high-frequency operation of a device. When neglecting the impact of the series resistances, the transit frequency can be approximated as

$$f_{\mathrm{T}} = \left[2\pi \left(\tau_{\mathrm{f}} + \frac{\sum C_n}{g_{\mathrm{m}}} \right) \right]^{-1}. \tag{2.2.2-3}$$

with $\sum C_n$ as the sum of all capacitances at the base node of the transistor. Since g_{m} is approximately proportional to the transfer current and the transit time τ_{f} remains reasonably constant before the onset of high-current effects, the transit frequency increases with increasing current.

Since in a HBT with $N_{\mathrm{B}} \gg N_{\mathrm{E}}$ the high-current effects are related only to the collector, not the base, the collector design has a major impact on the maximum f_{T}. The high-current region begins when the drift current in the collector is insufficient to carry the transfer current. Then, an electron pile-up at the BC interface follows, resulting in a diffusion current across the collector. The negative charge of the electrons is compensated by the injection of holes from the base into the collector. This is referred to as the Kirk effect.

The doping spike in the collector allows shaping the electric field so that the Kirk effect occurs at much higher current densities. It can be used in III-V HBTs since the epitaxial growth allows for excellent control over doping densities. In high-speed SiGe HBTs, where the collector doping is implanted and less fine control is possible, higher permissible current densities are commonly achieved by increasing the collector doping as a whole.

A result of the different doping and therefore space charge is an impact on the voltage dependence of the BC junction capacitance. The multiple regions of the collector doping need to be taken into account when modeling the capacitance. The material transition inside the collector similarly influences the relation of space charge to bias.

2.2.3 Top view and cross section

Fig. 2.3(a-c) shows the schematic top view and cross-sections of a typical III-V HBT. It consists of a number of layers that are commonly grown via epitaxy evenly on top of a semi-insulating substrate. The layers are then structured via (wet-)etching [UPR⁺11]. Very modern SiGe devices are grown in a similar manner with an epitaxial base (e.g. [FHB⁺11]), but more commonly, the cross-section is approximately as given in fig. 2.3(d).

Due to the fabrication technique, certain test structures cannot be used or must be adjusted for III-V HBTs, since it is not possible to remove a lower

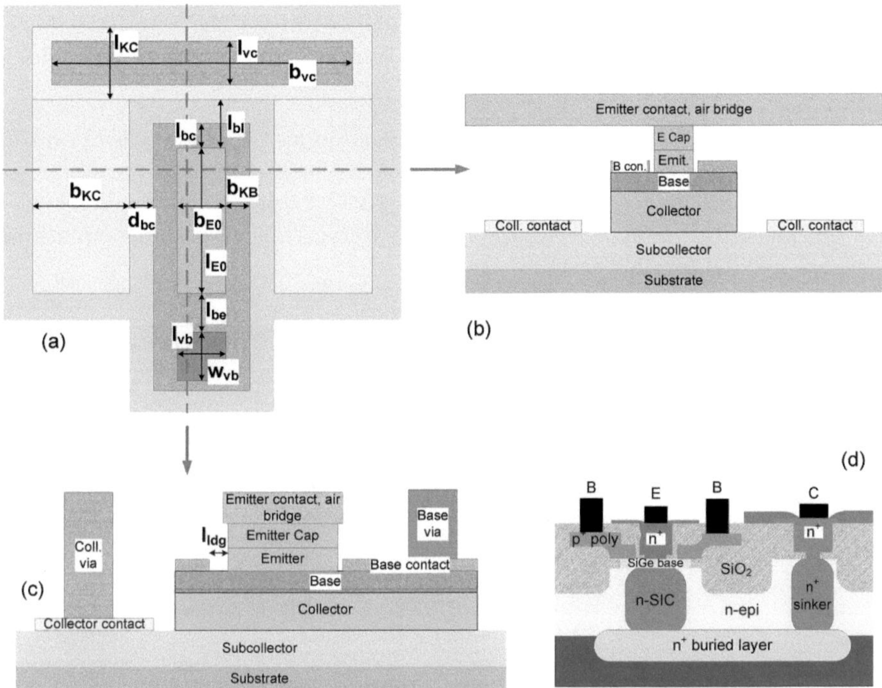

Figure 2.3: (a) Top view and (b,c) cross sections of a typical III-V mesa HBT structure. (d) Classical SiGe device structure for comparison. Modern SiGe HBTs can be fabricated in a mesa structure [RHF12]. All views are schematic. The emitter air bridge was removed from (a) for better visibility.

layer while keeping an upper layer. For example, the BE TLM structure requires a collector contact due to the existence of the BC diode, and implementing structures for the emitter resistance is not possible since the base cannot be omitted or changed.

Moreover, the mesa structure - named after the Latin word for table - can cause differences in the equivalent circuit:

- The "air bridge" style contact of the emitter for III-V devices, i.e. the emitter contact metal traversing the transistor, can result in a BE and CE parasitic capacitance.

- The very large external BC diode and therefore junction capacitance compared to SiGe HBTs, located directly under the semiconductor base, may require distributed modeling of the external collector and base resistances. Note, that attempts to reduce or remove the external capacitance by amorphization or etching in a transferred substrate technology have been made (e.g. [KML+08]).

- The base and emitter contacts are fabricated as a gold alloy layer on top of the semiconductor. While they can be seen as foreside contacts, it is easier to view them as surround resp. U-shaped contacts. While the former is common in SiGe, the latter is untypical. New scaling equations are needed for this case.

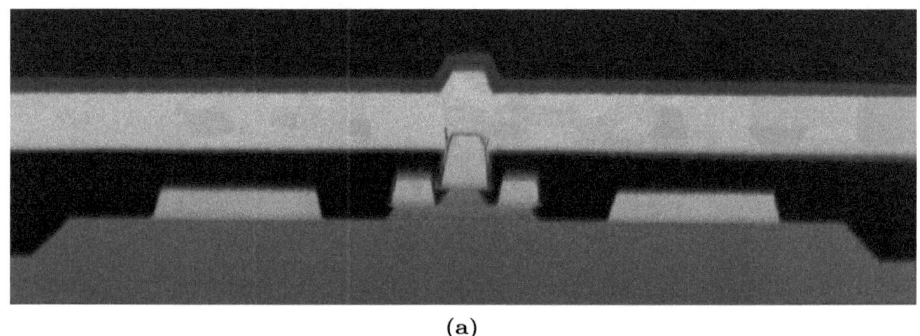

(a)

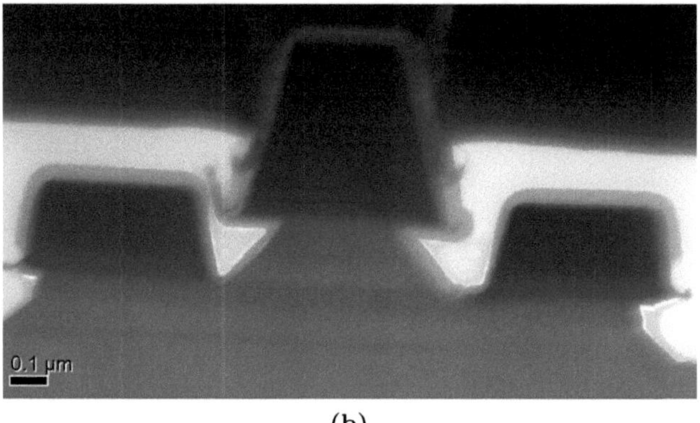

(b)

Figure 2.4: (a) SEM picture of InP HBT cross-section and (b) TEM picture of emitter contact and the BC mesa of an idential device.

For a better impression of a realistic III-V HBT shape, fig. 2.4 shows an SEM picture of a cross-section of the entire transistor and a TEM picture of the emitter and the BC mesa as well as the emitter and base contacts. In both pictures it is clearly visible that the rectangular shapes shown in fig. 2.3 are a simplification and trapezoidal shapes may be a better approximation for each mesa. This is important to note because it may lead to a bias-dependent effective emitter area as the space charge region width between base and emitter changes.

2.3 Compact modeling

2.3.1 Principles of compact modeling

2.3.1.1 Definition

In the context of this work, a compact model is an equivalent circuit consisting of one or more lumped elements which are described as a function of at least current and voltage. This work discusses models for bipolar transistors, but more simple devices such as diodes and even resistors can also be described by compact models. Compact models are primarily used to represent a complex device inside a circuit simulator.

Besides the physical basis as described below, a number of other criteria are relevant when choosing or developing a compact model. Foremost among them is numerical stability, i.e. successfully simulating the device under all conditions. Model run time is also important, especially when complicated circuits with dozens or hundreds of transistors are simulated. Finally, if applicable, a reliable and well-defined extraction sequence for model parameters is desirable. Therefore, most accurate model is not always the best model for a certain application.

2.3.1.2 Physical basis

Compact models for circuit simulators can be categorized on grounds of their physical basis:

- **Table-based models** interpolate the DC and RF behavior of the modeled device based on a grid of measured data. While very simple in principle, since little to no development effort is needed, these models cannot account for bias points outside of the measured range or indicate device properties that were not measured initially, such as noise. The results are not scalable and cannot be used for device analysis. Historically, these models could not account for large-signal effects, although modern measurement equipment like the Keysight PNA-X can create table-based models including these effects (e.g. [RHB$^+$08]).

  Despite the low complexity, table-based models can require a considerable measurement effort to create and maintain due to the large required measurement range and number of bias points. Inside the measurement range, however, they can be very accurate.

- **Empirical models** use arbitrary functions to describe measured data. Nearly as easy to create as table-based models, they may require a slightly lower measurement effort. However, they suffer from similar drawbacks: Extrapolation beyond the measured range may be erroneous, scalability is typically not included and there is no predictive modeling capability. Simultaneously, empirical models must be numerically stable, i.e. the equations used cannot be entirely arbitrary, but must be well conditioned for the implementation in a simulator.

- **Physics-based models** attempt to simplify the underlying differential equations for the device description sufficiently for an analytical description. At the same time, they maintain the link between the device structure and the electrical behavior through the use of an equivalent circuit whose elements are linked to device regions. The compact model parameters are based on the semiconductor material properties, geometrical and process-specific information, temperature, bias and physical constants.

  For this work, only physics-based compact modeling is of interest, since the other approaches are not suitable to improving the technology deployment of III-V HBTs.

Even in a physics-based model, not every equation is necessarily based directly on device physics. The model can contain empirical descriptions of certain physical effects, especially if the underlying math cannot be simplified or becomes to complicated to handle in a compact model. Similarly, if suitable solutions can only be found in certain operating regions, a smooth transition between regions must be assured, usually by means of an analytical transition function.

A physics-based model requires the most effort to create, since an understanding of the physics underlying the device is necessary, but is the most useful and versatile form of compact models. It enables predictive and statistical modeling and circuit design.

No model has a claim to represent all effects that can occur in a device, since all models are based on simplifications and neglect irrelevant influences for reasons of extraction effort and model run-time during simulations. However, as device dimensions, bias points and materials change, physical effects that were deemed irrelevant may become important. Compact models therefore require constant evaluation and development.

2.3.1.3 Scalability

One of the major advantages of physics-based compact models, with an equivalent circuit and model parameters that represent physical regions of a given device, is the extraction of a scalable model parameter set. Then, a user simply has to insert the device geometry and a suitable parameter set is automatically generated. This allows a circuit design to be optimized for specific applications or requirements. Additionally, it reduces the extraction effort, since not every device has to be measured at all bias points.

For bipolar transistors, there are often many options for a designer to choose from. Those can include

- the emitter stripe width and length

- the number of emitter, base and collector stripes

- the location of the collector contact relative to the emitter.

Under these circumstances, it is insufficient to include relatively simple scaling equations in the compact model, as is common for MOSFETs. Instead, a dedicated scaling tool can be used (e.g. TRADICA [SRR$^+$99]) that determines the concrete values of EC elements from the lateral geometry and technological data. The scaling of bipolar transistors is described in more detail in, e.g., [SC10].

However, scaling of a process makes demands not only on the compact model, but also on the technology; scaling (and modeling in general) is only meaningful for reproducible transistor characteristics. Additionally, verification by measurement is always needed. Given that InP foundries (and often III-V foundries in general) today typically do not offer fully scalable processes for their customers, but instead provide around three discrete device sizes (all with their own model cards) in a single contact configuration, scaling in this work is limited to the scaling of the emitter (resp. collector) area and perimeter. This is not uncommon for III-V HTBs; the FBH-model, for example, is formulated with current densities and the emitter area as parameters, but does not consider perimeter scaling [Rud06]. Parameter extraction including device scaling is demonstrated in section 4.5, with results shown in chapter 5.

2.3.2 The HICUM compact model

In this work, the HICUM compact model - short for <u>Hi</u>gh <u>Cu</u>rrent <u>M</u>odel - for bipolar transistors is used as a basis for modeling III-V HBTs. This section gives a brief overview over the model in order to aid understanding and the argumentation in this thesis. The interested reader is referred to more detailed descriptions in, e.g., [SC10, SPM15] for further reading.

HICUM has been in continuous development for over 30 years. It was initially developed to address the shortcomings of the well-known SPICE Gummel-Poon model. As the name implies, it was originally intended for modeling especially the high current region, in which f_T and f_{max} reach their maximum and which is therefore most suitable for high-speed operation. HICUM has been extended to include also high-frequency effects and correlated noise (e.g. [JDSC10, HSR$^+$12]). Additionally, HICUM is explicitly capable of handling band-gap differences at the BE junction [Paw14], and a conduction band spike forming at the base-collector (BC) junction as they occur not only in SiGe, but especially in III-V HBTs.

An additional advantage for modeling III-V HBTs is that HICUM offers multiple different model levels: The standard HICUM Level 2 (HICUM/L2), which includes all modeled physical effects and a separation between the internal and external transistors regions, and the less complicated, but also less physically accurate, HICUM Level 0 (HICUM/L0). The latter is a simplified variant that is more suitable for single-transistor parameter extraction due to a reduced parameter set and geometry separation and should be of interest for most III-V foundries. Other reasons for the use of HICUM/L0 are the faster run-time and the possibility to make a rough manual estimate of circuit performance based on model parameters before design and simulation, which is usually not possible for more complicated models.

Due to its more physics-based nature, this thesis focuses on HICUM/L2. Model extensions for III-V HBTs described in chapter 3 are, however, applicable to both model levels.

A third version of the compact model is HICUM Level 4 (HICUM/L4), which is intended for the simulation of distributed breakdown and self-heating. It consists essentially of multiple instances of HICUM/L2 or HICUM/L0 simulated in parallel and connected by parts of the base resistance. Each HICUM/L2 then represents a fraction of the transistor. HICUM/L4 is intended for very large transistors or very high power densities, when the assumption that a transistor region is represented by a single lumped element is no longer true. While this can be relevant especially for power amplifiers, so far, even a GaAs device with an emitter area of 2.2x20 µm^2 could be modeled satisfactorily with HICUM/L2. Therefore, HICUM/L4 is not discussed further in this thesis.

HICUM is supported as one of two standard BJT models by the Compact Model Coalition (CMC), a working group formed by certain companies in the semiconductor industry to pursue standardization and public access to compact models. This means that there is a current HICUM version available in all major circuit simulators and that the model code is available in Verilog-A; therefore, changes can be implemented quite easily.

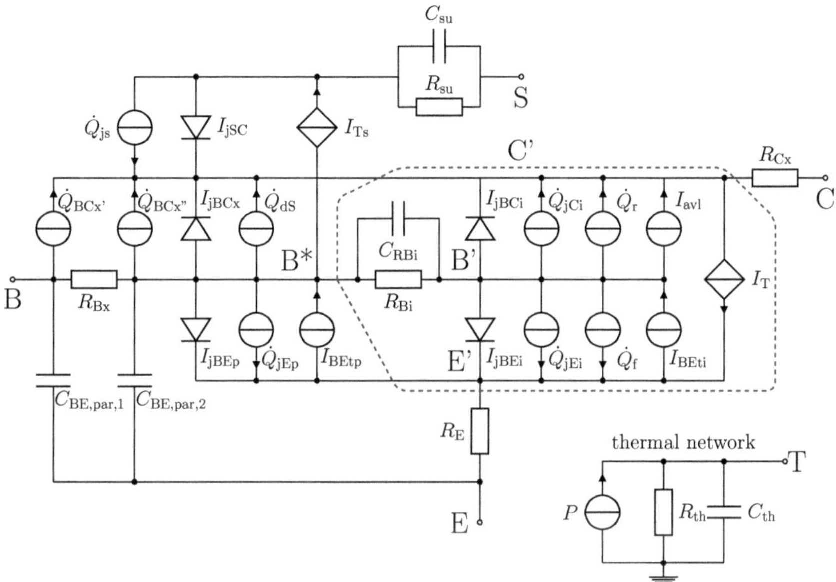

Figure 2.5: (a) Equivalent circuit of HICUM/L2 including the sub-circuit for self-heating. The sub-circuits for correlated noise and the non-quasi-static (NQS) effect are not shown. The dashed line indicates the internal transistors, i.e. the transistor relating to the area located vertically under the emitter window.

Fig. 2.5 shows the equivalent circuit of HICUM/L2. The adjunct networks for the vertical NQS effect and noise correlation were omitted here because neither was investigated in this work. The dashed line indicates the internal transistor, i.e. the ideal vertical transistor without the impact of external regions, external series resistances or the substrate. Removing the base resistance from the internal transistor essentially yields the (emitter-area-scaled) ideal 1D transistor that is commonly simulated in numerical device simulations, with the exception that the impact of the current spreading effect is still included.

2.3.3 Existing III-V compact models

Several dedicated III-V compact models exist in the literature and in practical use. Most notable among them are the UCSD model [UCS00] as the first generation of a III-V model that serves as the basis for subsequent iterations, the AHBT model [Agi16], which builds on the UCSD model and is supplied in the ADS circuit simulator, and the FBH model [Rud05] as the newest III-V specific compact model with built-in scaling capability. An overview over the models and their specific equations can be found in [Rud06]. While an in-depth investigation of each models capabilities is beyond the scope of this work, a brief discussion regarding the advantages of the concept presented in this work is warranted.

- **The UCSD model** is based on the SGPM. It contains formulations for a number of physical effects not accounted for in the basic model, among them self-heating, an equation for the 'soft-knee' current effect and an approach for the collector transit time. Despite a large number of improvements over its basis, though, it retains a number of problems common to models developed in this era. Among other issues, the diffusion charge in the model is formulated as a product of the transit time with the transfer current, which is incorrect and leads to a discrepancy between the model and measurements. While the BE- and BC currents are divided into internal and peripheral resp. external fractions, those diodes share a parameter set. The junction capacitance equations are not continuously differentiable. Thus, the model is not suitable for modern HBT applications anymore.

- **The AHBT model** improves on the shortcomings of the UCSD model by implementing the transit time formulation in a consistent way. Additionally, it includes an improved formulation for the junction charges and high-current transit time by borrowing the HICUM equations. However, in order to be compatible with UCSD model parameter sets, the DC formulation is mostly identical with the older model, especially containing a fixed reverse early voltage and identical BE- and BC diode currents. The formulation for the collector charge is purely empirical. To the best of the authors' knowledge, the model is not available in other simulators than ADS, and while the model equations are public, no official VA- or similar implementation exists.

- **The FBH model** is formulated independently of the UCSD and AHBT models. Originally intended for GaAs HBTs, it has been applied to InP HBTs as well. Its parameters intrinsically scale with the emitter area as a parameter. The model is intended to be relatively simple and easy to understand; it is, therefore, most comparable to HICUM/L0. The FBH model features a unique, but purely empirical formulation for the collector transit time. Equations for high-current effects are borrowed from HICUM. Given that HBT scaling is often more involved than simple multiplication with the emitter area, its inclusion into the model equation may in effect be detrimental.

In this work, the HICUM compact model is used as a basis and extended with new equations specifically addressing III-V-specific effects that are not included in the model. This approach has a number of advantages compared to the existing models:

- Most importantly, only parts of this work, especially regarding the parameter extraction, are specific to HICUM. In principle, the equations found in chapter 3 and the extraction methods discussed in section 4.3 are applicable to compact modeling in general.

- HICUM is a CMC standard model; its code is available as a VerilogA implementation and in all major circuit simulators. It features different model levels for maximum physical accuracy and parameter simplicity. Extending it for use with III-V semiconductors allows circuit designers used to SiGe devices to work in a familiar environment. This point becomes especially important with the possibility of integrating III-V devices on a silicon substrate.

- The transfer current formulation in HICUM is based on the GICCR, which addresses a number of relevant effects, especially injection across a heterojunction barrier, intrinsically in a physically meaningful manner. It uses the well-defined quasi-Fermi potentials directly at the contacts as inputs. Similarly, the transit time formulations cover, e.g., the impact of the SCR width variation on the base transit time. HICUM also includes a distributed and bias-dependent base resistance, independent diode parameter for the internal and external regions and non-quasi-static effects. It is therefore easier to include III-V specific effects in HICUM than attempt to improve other models.

- HICUM is in ongoing development. As new physical effects become more important through shrinking lateral or vertical dimensions, they are likely to be included.

- Other models already borrow HICUM equations, demonstrating their usefulness and acceptance in the HBT modeling community.

Therefore, this work contributes significantly to the quality of compact models available for III-V HBTs.

2.4 Numerical simulations of III-V materials

Numerical device simulations, i.e. the numerical solution of the differential equations governing the carrier behavior, is a valuable tool for compact modeling, since they permit a view into the device. Using this information, a device engineer can make assumptions about, e.g., the shape of the electric field in the collector with increasing current and draw appropriate conclusions for modeling.

Unfortunately, numerical simulation is very difficult when applied to III-V devices. The reasons for this are the NDM effect, which makes DD simulations impossible and HD simulations physically questionable, and the interface condition for an abrupt material changes. The Newton algorithm that such simulators use does not handle non-monotonous functions well and usually fails to converge when the NDM effect is relevant, which was verified by the technical support of a commercial simulator [Tik16]. The problem is not dependent on simulator; over the course of this work, multiple in-house (e.g. [Sch91b]) and commercial (e.g. [Syn15]) simulators were used with unsatisfactory results. Moreover, while I-V characteristics of III-V devices can regularly be found in literature, to the best of the author's knowledge, there are very few publications including AC or even quasi-static simulations. The few exceptions are either BTE simulations (e.g. [Roh02]) or HD simulations that are extensively calibrated using BTE simulations (e.g. [RPHJ^{+}07]), and are then only compared to measurements for a single value of V_{BC}. Therefore, it seems unlikely, that realistic III-V structures can be reliably simulated based on material parameters and a given doping and material profile. A similar conclusion was reached in [Wed16a], which focuses on the implementation of a deterministic BTE solver.

This section briefly summarizes simulation attempts for III-V structures.

2.4.1 Material properties

When simulating a semiconductor device, accuracy depends on many factors: sufficient numerical discretization, a well known doping and material profile, but also well known models and parameters for the semiconductor material to be simulated. Among the latter are models for the band-gap and relative band locations, but also for different recombination mechanisms, band-gap narrowing from high-doping effects, doping- and field-dependent mobility and others.

Silicon and Silicon-Germanium are well researched. When attempting to simulate a III-V device, though, a user may encounter uncommon materials for which no parameters exist. Furthermore, grading of the base and BC junction means that an arbitrary number of slightly different compounds must be handled.

Unfortunately, even commercial simulators such as SDEVICE [Syn15] do not supply appropriate parameters for all material models even for common III-V materials such as InGaAs lattice-matched to InP. Material parameter files are, however, accessible and the user is free to input other than the supplied values. Thus, a literature research was conducted for this work in order to find the required values to simulate type-I InP HBTs. A complete set of the parameters can be found in appendix A. This section presents the interpolation method used for the material systems [Ada82].

Whenever possible, material models were fitted to experimental data instead of taking parameter values for the models directly from literature. If no experimental data were available, data resulting from Monte-Carlo simulations as presented in the literature were used. It should be noted that the available data are not always consistent; especially in the case of parameters for the recombination model, the values found in the literature can vary by orders of magnitude. Absolute agreement between a measurement and a simulation can thus not be expected.

Material parameters P for arbitrary fractions x for a ternary material (i.e. consisting of three elements) AB_xC_{1-x} are typically interpolated using the polynomial

$$P(x) = a + b \cdot x + c \cdot x^2. \tag{2.4.1-4}$$

The parameters a, b and c are calculated from the known characteristics of the border materials AB and AC as well as one material with an arbitrary value for x [Ada82]. c is often referred to as a bowing parameter. An example for interpolating a ternary material is the use of known parameters for InAs and GaAs to form $In_{1-x}Ga_xAs$, which for $x = 0.47$ is lattice-matched to InP and is often used as the base in InP HBTs.

Parameters for quaternary materials of the form $A_{1-x}B_xC_yD_{1-y}$ can similarly be interpolated from the binary materials that make up their components

$$P(x,y) = (1-x)yP_{AC} + (1-x)(1-y)P_{AD} + xyP_{BC} + x(1-y)P_{BD}, \tag{2.4.1-5}$$

where P_{NM} is the parameter value of the material indicated by the indices. This format requires four initial data points and can become very imprecise. If no arbitrary values for x and y are required, it is often easier to use equation 2.4.1-4 instead of 2.4.1-5, using a ternary material as one of the borders B or C. An example for such an interpolation would be describing the quaternary $In_{1-x}Ga_xAs_yP_{1-y}$ with a fixed value of $x = 0.47y$ by using InP and $In_{0.53}Ga_{0.47}As$ as border materials. This material system is required to describe the base-collector grading in InP HBTs.

2.4.2 $n^+ - n - n^+$ structure

A good basic test for a numerical device simulation is an $n^+ - n - n^+$-structure. It is essentially a resistor, with the n^+-regions at the contacts preventing a direct impact of the contact boundary conditions on the simulation result. The doping profile for such a structure is given in fig. 2.6a. The semiconductor material for the structure is InP. Both the in-house simulator CHIEF and the commercial simulator SDEVICE were used. The profile is identical in both cases except for the discretization, since CHIEF runs 1D simulations, while SDEVICE always runs 2D simulations. The material parameters in CHIEF were chosen according to the literature research, those in SDEVICE according to the internal values except to move the velocity peak to $E = 14\mathrm{kV\,cm^{-1}}$.

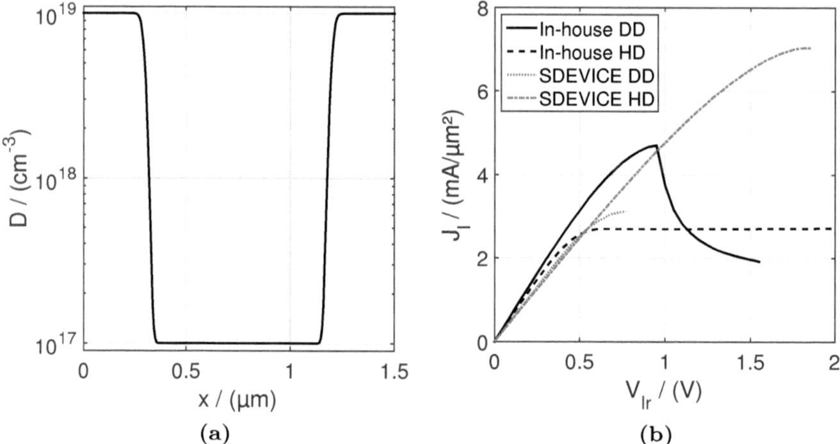

Figure 2.6: (a) Simulated doping profile and (b) I-V characteristic of the $n^+ - n - n^+$-structure for different simulators and solved equation systems.

The I-V characteristic shown in fig. 2.6b indicates that SDEVICE fails to converge when the carrier velocity reaches its maximum. Note, that the simulator can be made to converge if a set of material parameters with a sufficiently small difference between maximum and saturation velocity is chosen; since, however, the goal of numerical III-V simulations is in part to show the impact of the NDM effect on the transistor behavior, this is not a meaningful option. The Synopsis technical support verified complications when simulating the NDM model [Tik16].

Opposed to SDEVICE, CHIEF converges in HD mode because the formulation for the carrier driving force F_c was changed. In SDEVICE, and usually in the HD equation system, the driving force is calculated from the carrier temperature as [Syn15]

$$F_c = \sqrt{\frac{3k_B(T_c - T_l)}{2q\mu_c\tau_{e,c}}},\qquad (2.4.2\text{-}6)$$

with the carrier temperature T_c, the lattice temperature T_l, the carrier mobility μ_c and the energy relaxation time $\tau_{e,c}$. μ_c in turn depends on the driving force via

$$\mu_c(F_c) = \frac{\mu_{c0D} + v_{c,sat}\left(\frac{F_c^{\beta-1}}{E_{c0}^\beta}\right)}{1 + \left(\frac{F_c}{E_{c0}}\right)^\beta}.\qquad (2.4.2\text{-}7)$$

μ_{c0D} is the doping-dependent low-field carrier mobility, $v_{c,sat}$ is the saturation velocity and E_{c0} and β are model parameters. In SDEVICE, $\beta = 4$. This results in an interdependence of equations that is detrimental to the convergence of the simulator.

In CHIEF, this dependence was removed and the driving force formulated as

$$F_c = \frac{3k_B(T_c - T_l)}{2qv_{c,sat}\tau_{e,c}},\qquad (2.4.2\text{-}8)$$

While the simulator achieves convergence for nearly all investigated sets of material parameters, it is unclear how this change impacts the device behavior. A reference BTE simulation would be needed for comparison. A BTE simulator, in turn, would need to predict measurements with some accuracy in order to verify material models and parameters. No such simulator was available at the time of this writing. While a deterministic BTE simulator was developed in [Wed16a], significant differences between simulation and measurements still exist. A likely reason is that no interface condition for abrupt material transitions (e.g. [Sch90]) has been implemented so far. Additionally, the doping and material composition profiles for the investigated devices were not sufficiently well known.

The difference between the two formulations for the driving force and their impact on simulation convergence is illustrated in fig. 2.7 [Wed16b]. F_n was calculated based on (2.4.2-6) for a given carrier temperature (as is the case in the simulator), using a Newton algorithm to solve the coupled equation system and for a given mobility, solving (2.4.2-7) for F_c and (2.4.2-6) for T_c. In the latter case, a solution can be obtained for all mobility values, demonstrating that the $F_n - T_n$ characteristic is not unique. In the former case, the exact solution therefore depends on the initial conditions for the Newton algorithm. Using the low-field mobility results in a discontinuity for the driving force, though it is not known what exactly is implemented in any given simulator. For comparison, the characteristic is also shown for (2.4.2-8). The equation agrees well with the classical formulation for high carrier temperatures, but yields too high a driving force for low temperatures.

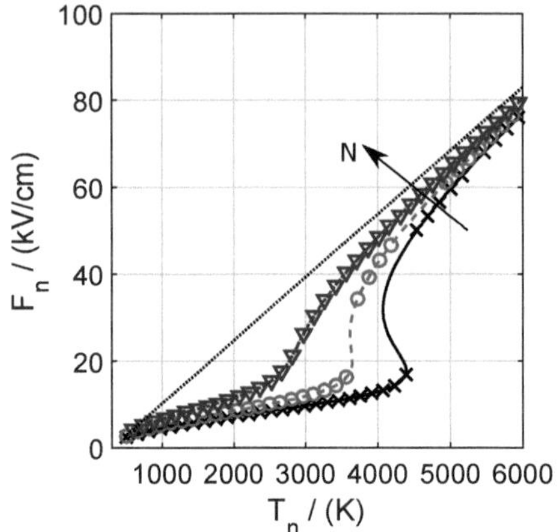

Figure 2.7: Driving force calculated from (2.4.2-6) for a given mobility (lines) and for a given carrier temperature (symbols) for different doping densities $N = [10^{16}, 10^{17}, 10^{18}]$ cm^{-3}, using the parameters for InP given in appendix A. The dotted line shows the driving force based on (2.4.2-8) for comparison.

2.4.3 HBT structure

Despite the uncertainties remaining for numerical simulations of III-V materials, a simulation of an HBT was attempted. A SIMS measurement of the doping profile for an HBT manufactured by the Fraunhofer Institut für angewandte Festkörperphysik (IAF) was available [DRM+11]. Since SIMS measurements suffer from some imprecision, it was used as a starting point for profile optimization; the goal was to achieve good agreement with the measured currents and transit frequency. A numerical quasi-static analysis and regional approach [ST06] were implemented in MATLAB for analysis.

The simulated doping and material profile is shown in fig. 2.8. In order to simulate both the base grading (material system GaAs to InAs) and the transition from emitter to base as well as base to collector (material system InP to $In_{0.53}Ga_{0.47}As$), both material systems were implemented in the simulator using the interpolation via bowing parameter according to (2.4.1-4).

The simulated doping profile is shown in fig. 2.8. It agrees closely with the SIMS results. During fitting to measured IV data, though, it was discovered that the transfer current depends almost entirely on an approximately 10 nm region around the BE junction; even small changes of doping or composition here can have a large impact on the transfer current. While this seems reasonable from a physics-based point-of-view (since the current through the device is dictated by the electrons overcoming the conduction band spike, the shape of which is governed by the doping and composition profile), it makes the SIMS measurements of the BE junction nearly useless due to insufficient precision. Note that the material parameters on the left and right side of the BE junction are identical; the apparent mole fraction spike seen in fig. 2.8b is due to the definition of the material systems in the simulator.

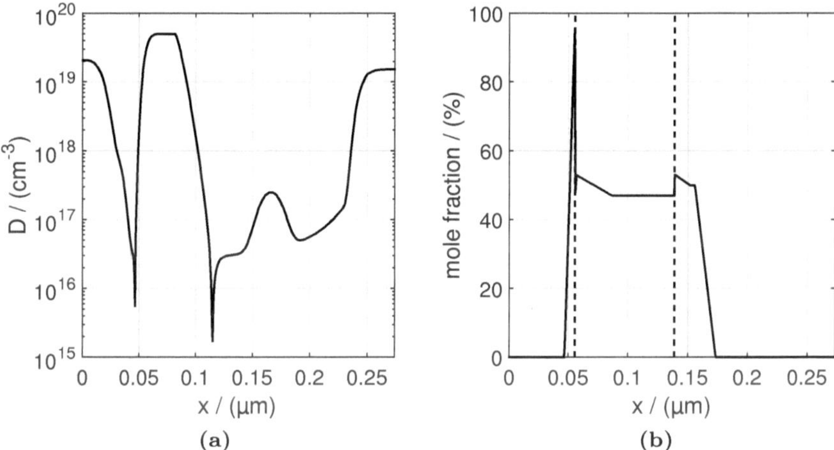

Figure 2.8: (a) Simulated doping profile and (b) simulated composition profile for InP HBT. Dashed lines in (b) indicate material system transitions.

Since a direct measurement of the internal transistor is impossible, but only 1D simulations were available, perfect agreement cannot be expected. From experience with scalable processes, though, the perimeter components of the transfer current and BE junction capacitance are nearly negligible. A comparison with the area-normalized properties serves well for an initial impression.

Fig. 2.9a shows the transfer characteristic of the simulation compared to measurements, normalized to the emitter area. Agreement is very good for $J_C > 0.01\,\mathrm{mA\,\mu m^{-2}}$. In the high current range, the difference is due to self-heating, which was not included in the simulator. The differences in the base current can be explained by the relatively large uncertainty for the recombination parameters and the fact that the simulation is one-dimensional, whereas the recombination current is a two-dimensional effect [TLB05].

The junction capacitances are a good indicator for rough agreement between simulated and measured profile. They are shown in fig. 2.9b. Agreement is very good, given that the actual and drawn areas may differ. Note, that the measured BC junction capacitance was normalized to the collector mesa area, not the emitter window area.

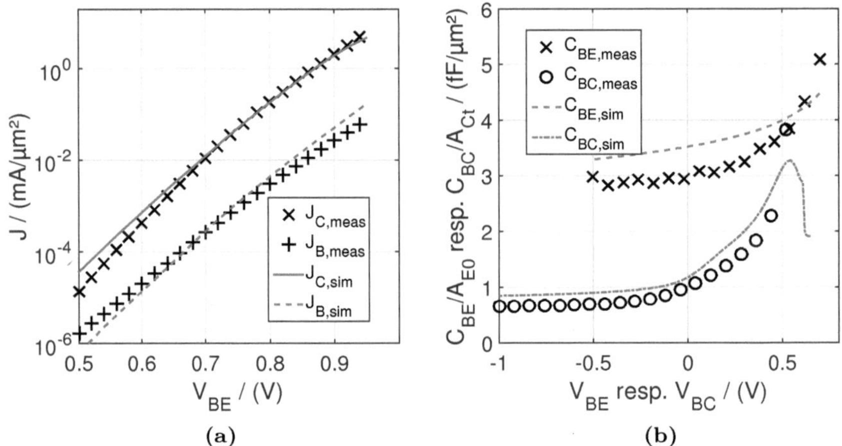

Figure 2.9: (a) Transfer characteristic at $V_{BC} = 0\,V$ and (b) junction capacitances for the simulated InP HBT.

The transit frequency, however, strongly differs between measurement and simulation. Given that the transistor is a 3D structure with a large BC area and parasitic influences, which were not removed from the measured data for a first comparison, a higher f_T is expected from the 1D simulation. However, as seen in fig. 2.10a, the simulated f_T is not only too low, but unlike the measurement shows hardly any dependence on V_{BC}, indicating a fundamental difference between measurement and simulation.

Analyzing the accumulated transit time in fig. 2.10b reveals three important regions:

- Nearly half the total transit time is related to the carrier overcoming the material transition from the emitter to the base. This is unrealistic; it is likely that a proper interface condition for the peak conduction band is needed.

- The rise of the transit time in the base is expected and corresponds to the neutral base transit time.

- At the BC junction, the transit time due to the BC SCR capacitance is visible. As approximately a third of the total transit time, it appears to be quite large, but bias-independent.

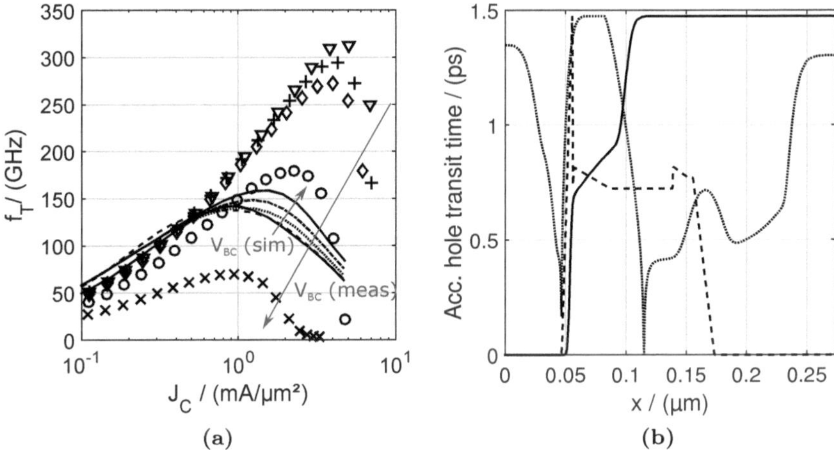

(a)   (b)

Figure 2.10: (a) Simulated (lines) and measured (symbols) transit frequency at $V_{\mathrm{BC}} = [-0.25, -0.1, 0, 0.25, 0.5]$ V and (b) accumulated transit time for $V_{\mathrm{BC}} = 0$ V and $f_{\mathrm{T}} = 0.8 \cdot f_{\mathrm{T,peak}}$. The dotted line in (b) indicates the doping profile, the dashed line the material composition.

2.4.4 Conclusions

Given that simulators show major convergence problems even for simple structures when the NDM effect is relevant unless changes for the driving force in HD simulations as in (2.4.2-8) are made and that simulations that include those changes show an unphysical charge behavior, numerical simulations of III-V materials are considered unreliable at best. Adding to this the many unknown material parameters, such as the important energy relaxation time, for many of the compounds that need to be simulated, and a distribution of the accumulated transit time that does not correspond well with theory, the physical accuracy of the simulations is uncertain.

While it is clearly possible to simulate III-V HBTs and achieve realistic values of J_{C} and f_{T} from previously published literature, this appears to be due to extensive calibration of material models and values on reference BTE simulations and measured data. From section 2.4.3, it is apparent that SIMS measurements are insufficient for this purpose.

Therefore, since the focus of this work is on compact modeling, not numerical device simulation, measured instead of simulated data are used as reference when current-dependent effects are modeled.

CHAPTER **3**

Model extensions

3.1 Introduction

Physics-based compact modeling requires a suitable equivalent circuit and a dedicated mathematical description for each relevant physical effect. Using HICUM as a starting point, an excellent basic model is already available. However, a number of additions must be made to accommodate the structure and properties of III-V HBTs as described in section 2.2. These changes are listed in this chapter.

While the temperature dependence in section 3.2 and the introduction of a parasitic capacitance in 3.3 are based on the device structure and physics, they require only minor changes to the model and are easily implemented thanks to the availability of the model Verilog-A code. The multi-region capacitance model in 3.4 and the formulation for the collector transit time in 3.5 require an in-depth derivation for physics-based modeling and are described in detail.

3.2 Temperature dependence of the base current components

In both HICUM and common III-V compact models, the base current components are modeled as diodes using the simple equation

$$I_{\mathrm{x}} = I_{\mathrm{xS}}(T) \left[\exp\left(\frac{V_{\mathrm{x}}}{m_{\mathrm{x}} V_{\mathrm{T}}} \right) - 1 \right], \qquad (3.2.0\text{-}1)$$

with the saturation current I_x and the nonideality factor m_x. The saturation current and the thermal voltage V_T depend on the ambient and actual device temperature. The former dependence is modeled as

$$I_{xS}(T) = I_{xS}(T_0) \left(\frac{T}{T_0}\right)^{\zeta_x} \exp\left[\frac{V_{gx}}{V_T}\left(\frac{T}{T_0} - 1\right)\right], \tag{3.2.0-2}$$

where I_{xS}, ζ_x and V_{gx} are placeholders to be filled in depending on the junction. The possible entries are listed in tab. 3.1.

Typically, each junction is modeled with two diodes with independent parameter sets. In SiGe, this represents the recombination current in the space charge regions in the low-current region and the injection current in the medium- and high-current region. In III-V HBTs, the distinction is less clear, since the base current consists primarily of the recombination current in the neutral base [RPHJ06, TLB05], though significant deviations can arise when using only a single diode equation for modeling the base current.

Likely reasons for different nonideality regions of the base current are, e.g., the different recombination mechanisms (surface vs. bulk recombination) or the impact of the current on the base transit time. It should be noted that, while it might be more physically accurate to model the base current via a recombination charge and the base transit time, there are multiple reasons to not attempt this: the lack of simulations to base accurate models on, the difficulty of describing both the spatial carrier concentration in the base and separating the base transit time from other components and the fact that using (3.2.0-1) works very well and is compatible with modeling of diodes and SiGe HBTs.

Table 3.1: Junction current components and temperature dependence parameters in HICUM/L2.

sat. current I_{xS}	bandgap V_{gx}	factor ζ_x
I_{BEiS}	V_{gE}	ζ_{BET}
I_{BEpS}	V_{gE}	ζ_{BET}
I_{REiS}	$\frac{1}{2}V_{gBE} = \frac{1}{2} \cdot \frac{V_{gE}+V_{gB}}{2}$	$\zeta_{REi} = m_g/m_{rei}$
I_{REpS}	$\frac{1}{2}V_{gBE} = \frac{1}{2} \cdot \frac{V_{gE}+V_{gB}}{2}$	$\zeta_{REi} = m_g/m_{rep}$
I_{BCiS}	V_{gC}	$\zeta_{BCi} = m_g + 1 - \zeta_{Ci}$
I_{BCxS}	V_{gC}	$\zeta_{BCx} = m_g + 1 - \zeta_{Cx}$

In HICUM, the temperature dependencies of the base current components are coupled to a number of other parameters, which is physically meaningful for a dominant injection current, but not for the recombination current. Additionally, some parameter value range restrictions need to be relaxed in order to capture the temperature dependence for III-V materials correctly.

Besides the currents, the parameters are also connected to the temperature dependence of

- the junction capacitances via the parameters $V_{\mathrm{gBE}} = \frac{V_{\mathrm{gE}}+V_{\mathrm{gB}}}{2}$ and $V_{\mathrm{gBC}} = \frac{V_{\mathrm{gC}}+V_{\mathrm{gB}}}{2}$ and m_{g},

- multiple parameters relating to the collector high-current transit time via ζ_{Ci}.

Especially the latter connection, if left in place, leads to large deviations for either the current or the transit time over temperature and must be changed.

When introducing new parameters, there is always a trade-off between extraction simplicity, model complexity, model accuracy and the physical reference. It is therefore not meaningful to simply make all possible temperature dependence parameters for the currents user-accessible. Instead, the following steps are taken for accurate compact modeling for III-V HBTs:

- The parameter value range limit for all ζ_{x} is removed.

- ζ_{BCi} and ζ_{BCx} are introduced as an independent parameter for both the internal and the external BC junction currents. It is extracted based on the BC base current. ζ_{Ci} is retained for the temperature dependence of the collector transit time parameters.

- V_{gBE} and V_{gBC} are introduced as user-accessible model parameters. They are extracted based on the low-bias base currents. The connection to the junction capacitances is retained since it remains physically meaningful; the band-gap voltages define the number of injected carriers and therefore influence the recombination rates.

3.3 Collector-emitter parasitic capacitance

Modeling the parasitic capacitances correctly is important not only for the correct scaling equations to be applied, but also for circuit-level application of the compact model, where the exact location of a capacitance in the EC is relevant.

The junction and parasitic capacitance components of HBTs are commonly extracted from so-called cold S-Parameter measurements, i.e. low-bias measurements during which the junction capacitances dominate the transistor behavior. Then, the impact of the junction diodes and the series resistances can be neglected because of the very low currents (at sufficiently low frequencies), the very simple equivalent circuit shown in fig. 4.31a can be used and the capacitances can easily be extracted from the Y-parameters. Since emitter and substrate are shorted, the capacitance C_{CE} is usually assumed to consist of only the CS junction capacitance. During III-V HBT measurements, however, it was observed that the capacitance is voltage-independent, which is a characteristic of parasitic capacitances. Due to the mesa design and the air-bridge contact used for the emitter, such a capacitance is physically realistic for III-V HBTs and was included in the HICUM equivalent circuit.

An example for the plate capacitor formed between the emitter and collector contacts is given in fig. 3.1. Assuming the measurements given therein and a relative permittivity of the benzocyclobutene (BCB) commonly used as an isolation and passivation material in III-V processes, the area-specific capacitance can be calculated as

$$\overline{C_{CEpar}} = \frac{\varepsilon_0 \varepsilon_r}{w_{EC}} \approx 0.046 \frac{fF}{\mu m^2}, \tag{3.3.0-3}$$

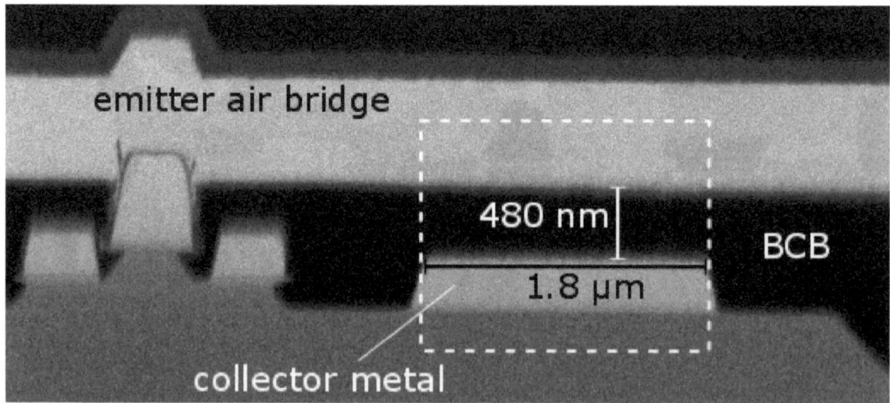

Figure 3.1: (a) Relevant measurements for the parasitic CE capacitance. The dashed rectangle marks the plate capacitor region between the collector and emitter contact metals.

with the distance between emitter air bridge and collector contact metal w_{EC}.

Measured data for the GCS process corresponding to the SEM figures in fig 3.1 is shown in fig. 3.2. For this process, the emitter air bridge length maximum is 5 µm, resulting in a total expected C_{CEpar} of 0.8 fF. This is clearly not the only contribution to C_{CE}, but the voltage-independence indicates that the remainder of the capacitance is also caused by the metallization, not the SC junction. The weak bias dependence seen in the figure for $V_{\mathrm{CE}} > 0.4V$ is due to the fact that the assumptions for the simplified equivalent circuit become invalid.

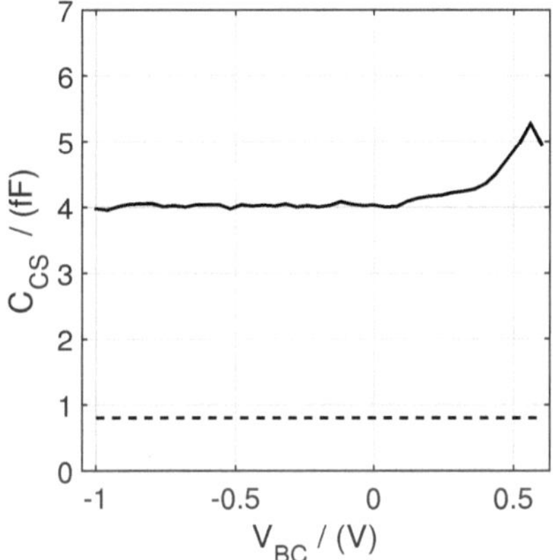

Figure 3.2: Measured $\mathrm{Im}(\underline{Y}_{21} + \underline{Y}_{22})\,/(2\pi f) = C_{\mathrm{CE}}$ of GCS process for $l_{\mathrm{E0}} = 15\,\mu\mathrm{m}$, $b_{\mathrm{E0}} = 0.8\,\mu\mathrm{m}$. The dashed line marks the value calculated from geometry.

3.4 Multi-region junction capacitance

The base-collector junction capacitance has a major impact on small- and large-signal HBT operation especially for mesa HBTs due to the large area of the junction. Attempts have been made to reduce its impact, such as the transferred-substrate approach [KML+08], but are not in use commercially and are unlikely to be applied at a large scale due to their technological difficulty and associated cost. With rising interest in low-power applications, accurate modeling of the transistor saturation region becomes more important [PSS+12]. Furthermore, even when transistors are operated under forward active conditions, large-signal bias swings require accurate models for the charges and capacitances under all bias conditions in order to properly represent the harmonic power levels.

III-V HBTs are often uniquely tailored towards certain applications by varying the doping profile and the material composition in the collector [RD15]. Due to the epitaxial layer growth, a high degree of control can be exerted over the doping concentration in different semiconductor layers, which is often used to include either a doping spike in the middle of the low-doped collector or a delta doping at the BC junction [RCC$^+$15] in order to prevent current blocking or to increase device linearity. This impacts the bias dependence of the junction capacitance.

Issues with the capacitance model are not neccessarily limited to III-V devices. Modern SiGe devices also do not feature a constant doping profile, but a strongly spatially dependent doping directly at the BC junction. In this case, the extracted parameters may lead to numerical problems in the transition functions, which need to be included for a smooth transition from punch-through across medium bias to the high bias region. Employing the multi-region model may lead to more physically accurate parameters.

The description of junction capacitances at the beginning of this work is explained as a basis in section 3.4.1. The derivation for the multi-region model is given in section 3.4.2. Voltage limiting and transition functions are briefly discussed in 3.4.3. The derived model is verified in 3.4.4, and the temperature dependence is shown in section 3.4.5.

3.4.1 Standard compact model capacitance description

This section summarizes the description of the junction capacitance at the beginning of this work. Note that all model parameters are named generically here, while the actual HICUM model parameters refer to a specific junction.

Modern HBT models are charge-based; furthermore, a VA implementation requires a charge formulation, since the capacitance are calculated as the derivatives of charges. Therefore, in order to describe the junction capacitance, the junction charge is calculated first from a solution of Poisson's equation using the total depletion approximation, i.e. the space charge is equal to the doping concentration, as

$$Q_{\mathrm{j}} = \frac{C_{\mathrm{j}0} V_{\mathrm{Dj}}}{1 - z} \left[1 - \left(1 - \frac{v_{\mathrm{j}}(V_{\mathrm{j}})}{V_{\mathrm{Dj}}} \right)^{1-z} \right], \qquad (3.4.1\text{-}4)$$

with the zero-bias capacitance $C_{\mathrm{j}0}$, the built-in potential V_{Dj} and the exponential coefficient z as model parameters and the limited voltage $v_{\mathrm{j}}(V_{\mathrm{j}})$, which is equal to the voltage across the junction V_{j} for $V_{\mathrm{j}} << V_{\mathrm{Dj}}$, and constant but smaller than V_{Dj} otherwise in order to prevent the numerical singularity at $v_{\mathrm{j}} = V_{\mathrm{Dj}}$. Note, that an analytical solution for $Q_{\mathrm{j}}(V_{\mathrm{j}})$ is only possible for simple doping profiles. Typically, spatially constant and linearly varying dopings are considered and the exponential parameter is added to compensate for the actual doping profile.

The junction capacitance is then determined from the differentiation of Q_{j}

w.r.t. V_j as

$$C_{j,\mathrm{CL}} = \frac{\mathrm{d}Q_j}{\mathrm{d}V_j} = \frac{C_{j0}}{\left(1 - \frac{v_j(V_j)}{V_{Dj}}\right)^z} \cdot \frac{\mathrm{d}v_j}{\mathrm{d}V_j}, \tag{3.4.1-5}$$

where the derivative $\frac{\mathrm{d}v_j}{\mathrm{d}V_j}$ forms a transition function which is equal to 1 for $v_j < V_j$ and smoothly tends to zero otherwise.

For large forward bias, the capacitance is kept constant by adding a linearly voltage-dependent charge

$$Q_{jf} = C_{jf}v_{jf}, \tag{3.4.1-6}$$

with the appropriate limited voltage v_{jf}.

For the BC capacitance, a simple punch-through case is usually included. Punch-through occurs when the BC SCR reaches the highly doped buried layer. Then, the doping on both sides of the SCR is high and the voltage dependence of the SCR extension is small. In VBIC [McA03] and the FHG HBT model [Rud06], the punch-through is included by adding a constant capacitance to the voltage-dependent part. In HICUM, a punch-through voltage is defined as a model parameter $V_{\mathrm{PTc(i,x)}}$ for the internal resp. external junction capacitance. For larger reverse bias, the capacitance transitions to a lower bias dependence with a fixed parameter $z_{\mathrm{C(i,x)r}} = 0.25 z_{\mathrm{C(i,x)}}$. The continuity of the capacitance is guaranteed by defining the punch-through zero-bias capacitance as

$$C_{\mathrm{jC(i,x)0,r}} = C_{\mathrm{jC(i,x)0}} \left(\frac{V_{\mathrm{DC(i,x)}}}{V_{\mathrm{PTc(i,x)}}}\right)^{z_{\mathrm{C(i,x)}} - z_{\mathrm{C(i,x)r}}} \tag{3.4.1-7}$$

and therefore

$$C_{\mathrm{jC(i,x),PT}} = \frac{C_{\mathrm{jC(i,x)0,r}}}{\left(1 - \frac{v_{j,\mathrm{PT}}(V_j)}{V_{Dj}}\right)^{z_{\mathrm{C(i,x)r}}}} \cdot \frac{\mathrm{d}v_{j,\mathrm{PT}}}{\mathrm{d}V_j}, \tag{3.4.1-8}$$

Note, that the punch-through voltage $V_{\mathrm{PTc(i,x)}}$ is defined as the internal collector voltage V_{ci} at punch-through and not the voltage across the junction. Therefore, the applied voltage at punch-through is calculated as

$$V_{\mathrm{jPT(i,x)}} = V_{\mathrm{PTj}} - V_{\mathrm{DC(i,x)}}. \tag{3.4.1-9}$$

The total junction capacitance is then calculated as the sum of the three contributions

$$C_{j,\mathrm{tot}} = C_{j,\mathrm{CL}} + C_{\mathrm{jC(i,x),PT}} + C_{jf}\frac{\mathrm{d}v_{jf}}{\mathrm{d}V_j}. \tag{3.4.1-10}$$

The total capacitance is continuous and smooth because the derivatives of the limited voltages supply transition functions with values between zero and one. The use of transition functions resp. voltage limiting is discussed in section 3.4.3.

3.4.2 Derivation for the multi-region model

In the presence of doping steps and material layer stacks, a single-region model is insufficient to describe the junction capacitance. This becomes evident from fig. 3.3 showing schematically the electric field in a single material collector layer for three different doping concentrations N_1, N_2 and N_3. For material steps, the difference in their relative permittivity ε_{ri} causes discontinuities of the electric field according to the displacement field continuity

$$\varepsilon_{r1} E_1 = \varepsilon_{r2} E_2 \tag{3.4.2-11}$$

at the transition points. Additionally, the differences in the bandgap of two semiconductor materials directly affect the built-in voltages of the junctions, which further depend on the doping concentrations on either side of the junction. In the multi-region model, this is taken into account by assigning a built-in voltage V_{Dk} for each region k.

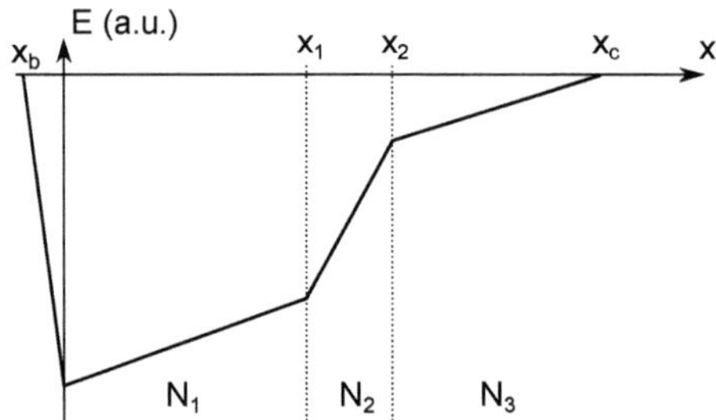

Figure 3.3: Schematic representation of the electric field in a single material collector for three doping regions. x_1 and x_2 mark the region boundaries, x_c and x_b the extensions of the SCR into the collector and base, respectively.

The following assumptions are made for this derivation:

- Total depletion approximation, i.e. the space charge is given by the doping concentration $\rho_k = q \cdot N_k$.

- The base doping concentration N_B is much larger than the collector doping concentration and therefore the extension of the SCR into the base is nearly zero and is neglected.

- The doping in a region k is spatially constant.

Furthermore, for the sake of simplicity, only the variation of the doping is considered here. The equations become significantly more complicated when the discontinuity of the relative permittivity is taken into account and are briefly demonstrated in appendix B.

The area-specific junction charge can be calculated as

$$\overline{Q_j} = \sum_{i=1}^{n} qN_i\left(x_i - x_{i-1}\right),\qquad(3.4.2\text{-}12)$$

with the region number i, the region extensions into the collector x_i, the BC junction $x_0 = 0$ and the total SCR extension $x_n = x_c$. The capacitance is determined by taking the derivative w.r.t. the junction voltage V_j, keeping in mind that only x_n is voltage dependent, resulting in

$$\overline{C_j} = \frac{\mathrm{d}\overline{Q_j}}{\mathrm{d}V_j} = \frac{\mathrm{d}\overline{Q_j}}{\mathrm{d}x_n}\frac{\mathrm{d}x_n}{\mathrm{d}V_j} = qN_i\frac{\mathrm{d}x_n}{\mathrm{d}V_j}.\qquad(3.4.2\text{-}13)$$

Solving the one-dimensional Poisson equation for constant relative permittivities

$$\frac{\mathrm{d}^2\psi}{\mathrm{d}x^2} = -\frac{\rho}{\varepsilon_0\varepsilon_r}\qquad(3.4.2\text{-}14)$$

yields the voltage dependent extension of the SCR for two regions

$$w_{SCR} = x_c = \sqrt{\frac{2\varepsilon(V_{D,2} - V_j)}{N_2 q} + \frac{N_2 - N_1}{N_2}\cdot x_1^2}.\qquad(3.4.2\text{-}15)$$

with the permittivity $\varepsilon = \varepsilon_r\varepsilon_0$. V_{Di} is the built-in voltage of the ith region from the BC interface without any additions. Adding a third region yields

$$w_{SCR} = \sqrt{\frac{2\varepsilon(V_{D,3} - V_j)}{N_3 q} + \frac{N_3 - N_2}{N_3}\cdot x_2^2 + \frac{N_2 - N_1}{N_3}\cdot x_1^2}.\qquad(3.4.2\text{-}16)$$

The impact of the additional regions can be included in an effective built-in potential and written for the general multi-region case as

$$V_{D,k,\mathrm{eff}} = V_{D,k} + \frac{q}{2\varepsilon}\sum_{i=1}^{k-1}(N_{i+1} - N_i)x_i^2.\qquad(3.4.2\text{-}17)$$

With the effective built-in voltage, the SCR width becomes

$$w_{\text{SCR}} = \sqrt{\frac{2\varepsilon(V_{\text{D},k,\text{eff}} - V_{\text{j}})}{N_k q}}, \tag{3.4.2-18}$$

i.e. the capacitance in each region can be described by the standard capacitance equation, but the doping in regions closer to the BC interface affect the effective built-in voltage.

Formulating the charge in terms of the voltage instead of the x-coordinate as required for a compact model yields

$$Q_{\text{j}} = C_{\text{jf}} \cdot (v_0 - V_{\text{tr},0}) - Q_{\text{j0}} +$$
$$\sum_{i=1}^{n} \frac{C_{\text{j0}i} V_{\text{D},i,\text{eff}}}{1 - z_i} \left[1 - \left(1 - \frac{v_i(V_j)}{V_{\text{D},i,\text{eff}}} \right)^{1-z_i} \right]. \tag{3.4.2-19}$$

with the basic model parameters $C_{\text{j0}i}$, $V_{\text{D}i}$, z_i, the limited voltages v_i, the forward-bias constant capacitance value C_{jf} in analogy to (3.4.1-6) and the zero-bias correction charge

$$Q_{\text{j0}} = \sum_{i=1}^{n} \frac{C_{\text{j0}i} V_{\text{D},i,\text{eff}}}{1 - z_i} \left[1 - \left(1 - \frac{v_i(V_j = 0)}{V_{\text{D},i,\text{eff}}} \right)^{1-z_i} \right]. \tag{3.4.2-20}$$

required due to the the influence of the limited voltages and included so that $Q_{\text{j}}(V_{\text{j}} = 0) = 0$.

The transition from one region (k) to the next one $(k+1)$ occurs when the region closer to the junction becomes fully depleted and the SCR boundary starts encroaching region $(k + 1)$. The corresponding transition voltage is given by

$$V_{\text{TR}k} = V_{\text{D},k} - \sum_{i=1}^{k} \frac{q N_i}{\varepsilon} \left[\frac{w_i^2}{2} + w_i x_{i-1} \right]. \tag{3.4.2-21}$$

where w_k is the width of region k. The depletion approximation with abrupt doping profile changes causes discontinuous derivatives at the transition between regions. In order to reflect realistic (smooth) doping transitions and hence to avoid the discontinuities, the transitions are smoothed by the transition functions.

While different transition and voltage limiting functions are discussed in section 3.4.3, the practical implementation of the model derived up to this point follows the methodology in HICUM and uses the same voltage limiting. Generally, the smooth transitions between adjacent voltage (and layer) regions are accomplished by

$$v_k(V_j) = V_{tr,k-1} + a_k \ln\left[1 + \exp\left(\frac{V_{tr,k} - V_j}{a_k}\right)\right] - $$
$$a_{k-1} \ln\left[1 + \exp\left(\frac{V_{tr,k-1} - V_j}{a_{k-1}}\right)\right], \tag{3.4.2-22}$$

i.e. v_k equals V_j in region k as long as v_k is sufficiently far from the transition voltages. When reaching either region boundary, v_k starts to become limited to the respective transition voltage. For the last region n, a transition only from the previous region $n-1$ is necessary, so that limiting needs to be applied only towards $V_{tr,n-1}$:

$$v_n = V_{tr,n-1} - a_{n-1} \ln\left[1 + \exp\left(\frac{V_{tr,n-1} - V_j}{a_{n-1}}\right)\right]. \tag{3.4.2-23}$$

At high forward bias the depletion capacitance reaches a peak and then drops to zero over a certain voltage range, since the depletion charge has to remain finite. In practice, this drop cannot be measured since it is masked by the much larger diffusion capacitance. Therefore, the computationally most simple approach is to replace V_j by the auxiliary voltage

$$v_0 = V_{tr,0} + a_0 \ln\left[1 + \exp\left(\frac{V_j - V_{tr,0}}{a_0}\right)\right]. \tag{3.4.2-24}$$

$V_{tr,0}$ can either be calculated from the measured ratio C_{jf}/C_{j01} (like in HICUM) or used directly as model parameter. Note that in the discussion above $V_{tr,0} > 0$ holds while $V_{tr,k}$ is the kth punch-through voltage, which is generally less than zero. The parameters a_k, which determine the width of the transition regions, can be made accessible to maximize the accuracy. Finally, from the necessary continuity of the $C_j(V_j)$ relation, the zero-bias capacitance of the region k can be calculated as

$$C_{j0k} = C_{j01} \cdot \prod_{i=1}^{k-1} \frac{(1 - V_{TR,i}/V_{D,i+1,\text{eff}})^{z_{i+1}}}{(1 - V_{TR,i}/V_{D,i,\text{eff}})^{z_i}}. \tag{3.4.2-25}$$

3.4.3 Voltage limiting

Due to the requirement of continuously differentiable charge and current equations for circuit simulation, a transition function must be employed in the capacitance model that provides a continuous crossover from one region to the next. The transition function in the capacitance terms arises directly as a consequence of using the limited voltage in the charge calculation as in (3.4.1-5). Therefore, a transition function must fulfill the following conditions:

- Its values must be between zero and one for arbitrary inputs.

- It needs to be integrable.

- It needs to be continuously and infinitely differentiable.

- Its influence on adjacent regions should fade away quickly.

Given these conditions, several analytical functions are investigated. Exponential limiting in the form of

$$T_{\mathrm{exp}} = \frac{\exp\left(\frac{V_{\mathrm{TR}} - V_{\mathrm{j}}}{A_{\mathrm{exp}}}\right)}{1 + \exp\left(\frac{V_{\mathrm{TR}} - V_{\mathrm{j}}}{A_{\mathrm{exp}}}\right)} \qquad (3.4.3\text{-}26)$$

is used in many compact models such as HICUM and MEXTRAM. Square root limiting

$$T_{\mathrm{sqrt}} = \frac{1}{2}\left(1 - \frac{V_{\mathrm{j}} - V_{\mathrm{TR}}}{\sqrt{(V_{\mathrm{j}} - V_{\mathrm{TR}})^2 + A_{\mathrm{sqrt}}}}\right) \qquad (3.4.3\text{-}27)$$

is used in VBIC. Trigonometric functions as in

$$T_{\mathrm{atan}} = \frac{\frac{\pi}{2} - \arctan[A_{\mathrm{atan}}(V_{\mathrm{j}} - V_{\mathrm{TR}})]}{\pi} \qquad (3.4.3\text{-}28)$$

or hyperbolic functions such as

$$T_{\mathrm{tanh}} = \frac{1 - \tanh\left[A_{\mathrm{tanh}}(V_{\mathrm{j}} - V_{\mathrm{TR}})\right]}{2} \qquad (3.4.3\text{-}29)$$

are also natural choices for transition functions. Upon closer examination, however, (3.4.3-29) can be rewritten in terms of exponential functions as

$$T_{\mathrm{tanh}}(x) = \frac{1}{2}\left[1 - \tanh(x)\right] = \frac{e^{-x}}{e^{x} + e^{-x}} = \frac{e^{-2x}}{1 + e^{-2x}}, \qquad (3.4.3\text{-}30)$$

where $x = A_{\text{tanh}}(V_j - V_{\text{TR}})$. Therefore, (3.4.3-30) is identical to (3.4.3-26) for $A_{\text{tanh}} = A_{\text{exp}}/2$ and will not be considered independently.

Integration of (3.4.3-26) - (3.4.3-28) leads to the limited voltage functions

$$v_{\text{exp}} = V_{\text{TR}} - A_{\text{exp}} \ln \left[1 + \exp \left(\frac{V_{\text{TR}} - V_j}{A_{\text{exp}}} \right) \right], \qquad (3.4.3\text{-}31)$$

$$v_{\text{sqrt}} = V_{\text{TR}} + \frac{1}{2} \left(V_j - V_{\text{TR}} - \sqrt{(V_j - V_{\text{TR}})^2 - A_{\text{sqrt}}} \right) \text{ and} \qquad (3.4.3\text{-}32)$$

$$v_{\text{atan}} = \frac{\pi A_{\text{atan}}(V_j + V_{\text{TR}}) + \ln \left[A_{\text{atan}}^2(V_{\text{TR}} - V_j)^2 + 1 \right]}{2\pi A_{\text{atan}}} + \frac{2 A_{\text{atan}}(V_j - V_{\text{TR}}) \arctan \left[A_{\text{atan}}(V_{\text{TR}} - V_j) \right]}{2\pi A_{\text{atan}}}, \qquad (3.4.3\text{-}33)$$

respectively.

For this comparison, a transition parameter A has been included in all functions. It was arbitrarily chosen as equal to the thermal voltage V_{T} for the exponential transition function and fitted for closest possible agreement for the two other functions. In practice, this parameter is usually set to either a fixed value or a fixed fraction of the built-in potential [SC10, Gil10, MSB$^+$96]. In order to investigate only the impact of the transition functions, the capacitance equation parameters are identical on both sides of the transition.

As seen in fig. 3.4a and fig. 3.4b, the atan-function deviates slightly from a value of one for a wide region and results in an offset from Vj. Evaluation of the limited voltage reveals that it is not, in fact, limited and tends to infinity for infinite input voltages due to the logarithm in (3.4.3-33). This disqualifies the atan-function as possible transition function since it may lead to numerical instabilities. The effect is also visible in the normalized capacitance shown in fig. 3.4c.

The impact of the transition functions on the capacitance is most visible in the derivatives, which are responsible for higher-order harmonics. The second derivative of the capacitance w.r.t. the applied voltage is shown as an example in fig. 3.4d. While the square root function has a slightly narrower spike in the second derivative than the exponential transition, the value of its transition function is different from one for a much larger range than the exp-function. Therefore, the exponential transition is most suitable.

The computational effort of the transition function was not investigated here. While it is expected that the evaluation of the exponential transition is slightly slower, its better performance justifies a small trade off in speed.

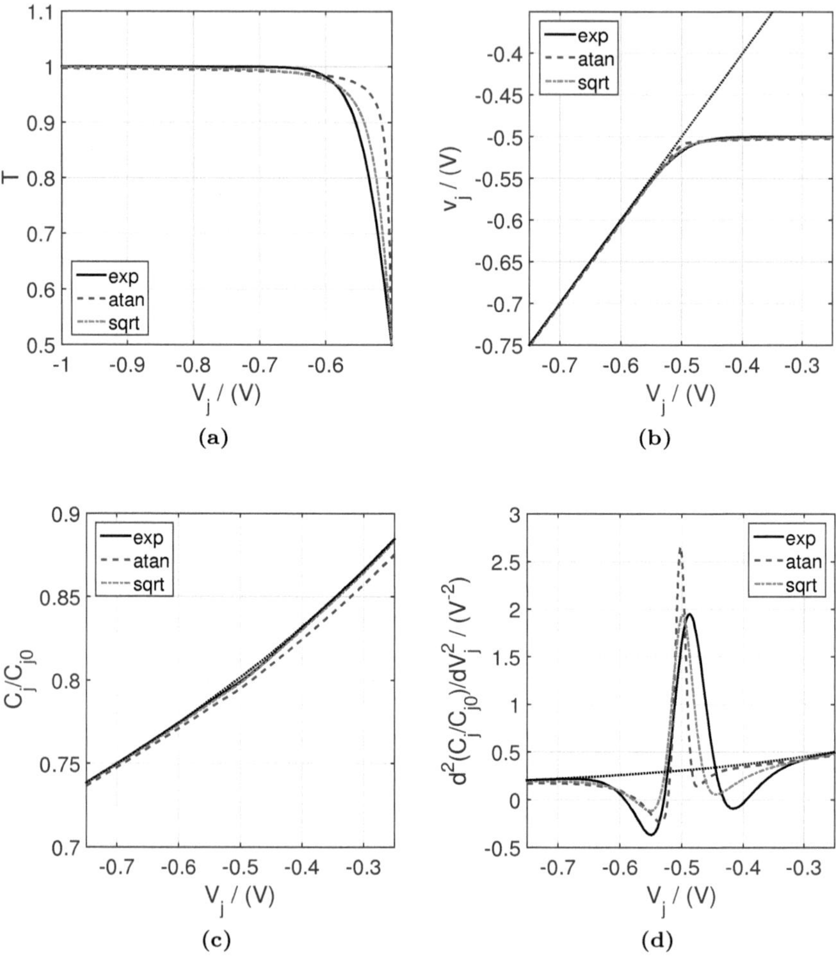

Figure 3.4: Comparison of transition functions properties versus junction voltage. (a) transition function value (half of symmetrical function shown); (b) voltage limiting; (c) transition between identical capacitance regions; (d) second derivative of capacitance including transition. Dotted line in (b) - (d): characteristic without limitation or transition. The transition voltage is $V_j = -0.5\,\mathrm{V}$.

3.4.4   Model verification

In the theoretical derivation, total depletion within the SCR and no extension of the SCR into the base were assumed. Since neither assumption is true, the model is verified here based on a numerical solution from an in-house drift-diffusion device simulator [Sch91b] and measurements of fabricated devices.

Three doping profiles have been chosen for this comparison: (1) a long, lightly-doped collector typical for high-voltage HBTs as a reference; (2) a lightly-doped collector with a doping spike in the center (e.g., [ZKC$^+$12]); (3) a lightly-doped collector with a doping spike directly at the BC junction. All profiles also contain a highly doped buried layer/subcollector. The simulation results are shown in fig. 3.5 in comparison with three model types: a single-region model (3.4.1-5), a single region model including a simple punch-through model (3.4.1-10), as described in section 3.4.1, and the multi-region model in which all parameters for the respective regions are accessible (3.4.2-19). Additionally, fig. 3.5d presents a comparison of these models with measured data from a high-speed InP DHBT (cf. also section 5.2).

The improvement gained by using a multi-region model is clear for all cases except profile 1. Note that, even though profile 2 is a three-region profile, the region closest to the junction is very short and low-doped; the resulting capacitance can thus be modeled with a two-region model.

The relative error of the classical model with punch-through can exceed 20% in the relevant bias region. Equally important for large-signal models is that the slope of the two-region model curve correlates well with both the device simulation and experimental results, avoiding errors in the higher-order derivatives. Furthermore, the model parameters remain physics-based and thus enable statistical modeling.

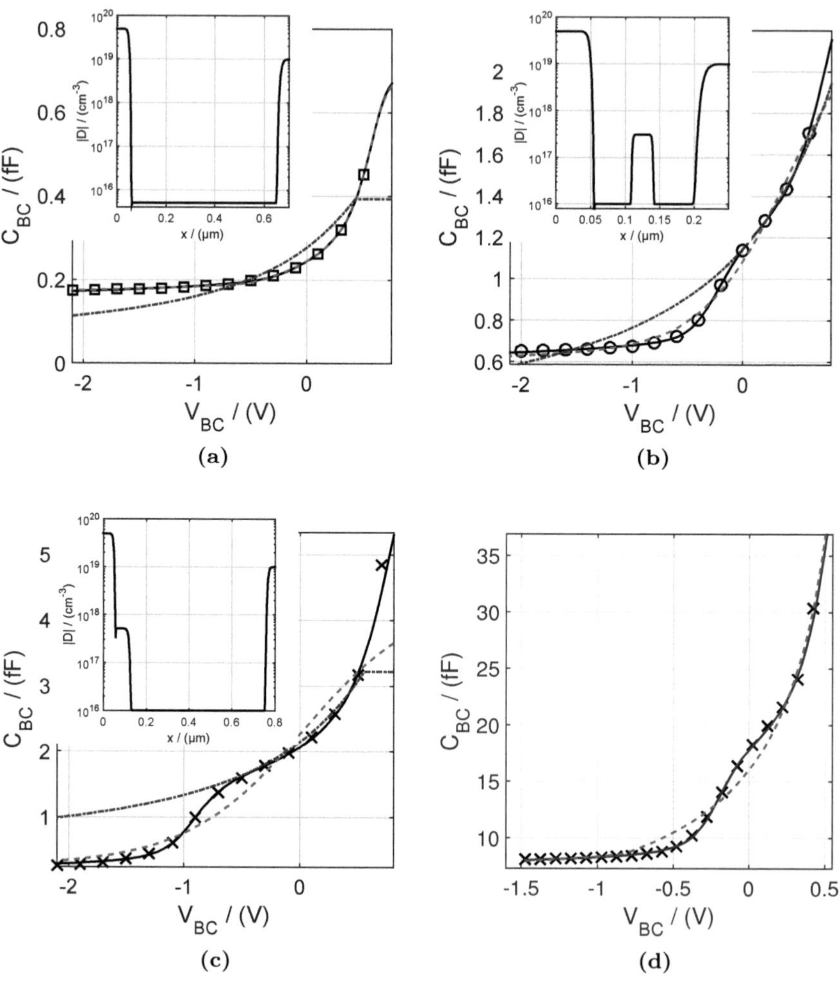

Figure 3.5: Simulated resp. measured junction capacitances (symbols), single-region model with simple punch-through (dashed line), single-region model without punch-through (dash-dotted line, figs. (a-c)) and multi-region model (solid line). (a) Profile 1, (b) profile 2, (c) profile 3; the insets show the simulated doping profiles. (d) Measured junction capacitance of an InP DHBT (x) vs. models .

3.4.5 Temperature dependence

Given complete ionization, C_{j0} is only temperature-dependent via V_{Dj}, which in turn is temperature-dependent via the intrinsic carrier density and possible band-gap changes in multi-material systems. Then, the temperature dependence for a single region is

$$C_{j0}(T) = C_{j0}(T_0) \left(\frac{V_{Dj}(T_0)}{V_{Dj}(T)} \right)^z \text{ and} \qquad (3.4.5\text{-}34)$$

$$V_{Dj}(T) = V_{Dj}(T_0) \left(\frac{T}{T_0} \right) - m_g V_T \ln \left(\frac{T}{T_0} \right) - V_g(T_0) \left(\frac{T}{T_0} - 1 \right), \quad (3.4.5\text{-}35)$$

with the bandgap voltage V_g and the temperature dependence parameter of the bandgap m_g, both used as model parameters. (3.4.5-35) is the temperature dependence used in HICUM except for the limiting functions, which can be neglected for $V_{Dj}(T) >> V_T$ [SC10].

While in theory each region should have its own temperature parameter set, during applications of the multi-region formulation, it has turned out that using a single set is sufficient for describing the temperature dependence. Additionally, the extraction effort is too high and the cold S-parameter measurement accuracy too low to justify the introduction of additional temperature parameters.

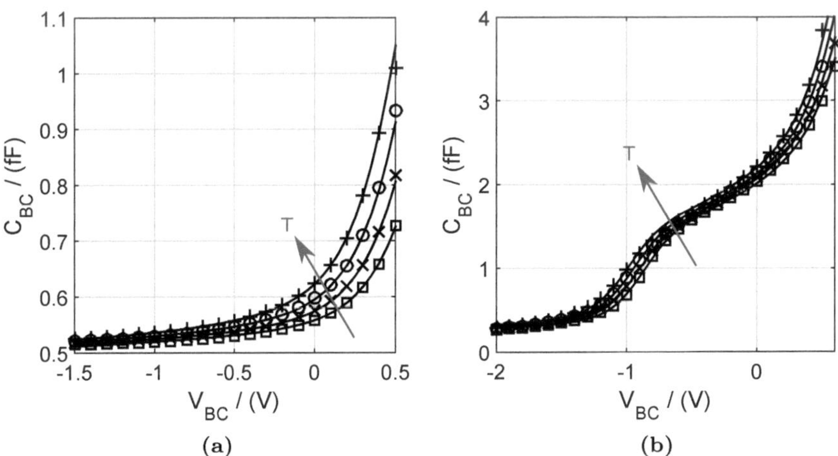

Figure 3.6: Junction capacitance from simulation (symbols) and two-region model (lines) of (a) profile 1 and (b) profile 3 for $T =$[250, 300, 350, 400] K.

The temperature dependence of the built-in voltage affects the transition voltages as well, because it changes the terminal voltage at which a region is fully depleted (c.f. (3.4.2-21)), and is included in the model as

$$\Delta V_{\mathrm{TR},k} = -\Delta V_{\mathrm{Dj},k}.$$ (3.4.5-36)

Fig. 3.6 shows a comparison of the device simulation result and the two-region model for four different ambient temperatures. Fig. 3.6a uses profile 1, the single-region profile with a simple punch-through, as a reference for the achievable accuracy while fig. 3.6b shows the data relating to profile 3 to demonstrate the impact of additional regions. Good agreement is obtained in the region of interest for a single temperature parameter set.

3.5 Collector transit time

The negative differential mobility of III-V materials causes a behavior of the carriers that is fundamentally different from that in SiGe for high electric fields: as the field strength increases, the carriers gain energy and transition from the Γ- to the L-valley, causing a decrease in carrier velocity. Therefore, with increasing field strength, the collector transit time increases in III-V materials where it would decrease or remain constant in SiGe.

In a well-designed HBT, the only region in which field strength is sufficiently large for the NDM effect to be relevant is the collector. Simultaneously, the field distribution in the collector is current dependent, since the electrons injected into the collector compensate the space charge of the doping.

This section is divided into two parts. In the first part, the low-current collector transit time is derived from physical principles. The considerations here can are also applicable to SiGe HBTs and can be included in SiGe compact models to improve the transit time formulation. In the second part, multiple approaches for the current-dependent modeling are compared and evaluated. This section is unique to III-V HBTs, since, due to the NDM effect, there is a medium-current region in which the collector transit time decreases before the onset of high-current effects.

3.5.1 Low-current transit time

In this section, it is assumed that neither the BC SCR width w_{BC} nor the carrier velocity $v(|E|)$ or the electric field $|E(x)|$ depend on the transfer current density.

The base-collector transit time is given by [SC10]

$$\tau_{\text{BCv}} = \int_0^{w_{\text{BC}}} \left(1 - \frac{x}{w_{\text{BC}}}\right) \frac{1}{v(|E(x)|)} dx. \tag{3.5.1-37}$$

Therefore, in order to find an analytical description of the transit time, a relation of the velocity to the electric field as well as the distribution of the electric field in the collector are required.

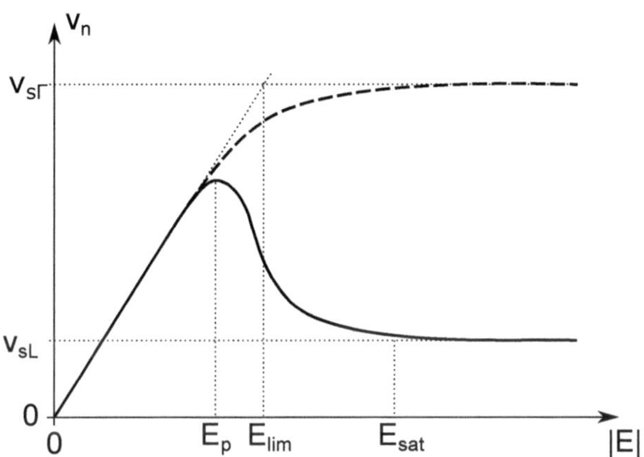

Figure 3.7: Schematic velocity-field characteristic for III-V materials including the relevant parameters for the equations in this chapter: the electric field at peak velocity E_{p}, the Γ-valley saturation field E_{lim} and the overall saturation field E_{sat}. Note that E_{sat} is not well defined and used only qualitatively.

3.5.1.1 Inverse velocity model

For a single conduction band, the velocity-field relation can be written as

$$v_{\text{n}} = v_{\text{s}} \frac{u}{(1 + u^\beta)^{1/\beta}}, \tag{3.5.1-38}$$

with the saturation velocity v_{s} of the valley and the normalized field

$$u = \frac{|E|}{E_{\text{lim}}}, \tag{3.5.1-39}$$

where $E_{\text{lim}} = v_{\text{s}}/\mu_{\text{0D}}$. μ_{0D} is the doping-dependent low-field carrier mobility

$$\mu_{\text{0D}} = \mu_{\text{min}} + \frac{\mu_{\text{max}} - \mu_{\text{min}}}{1 + \left(\frac{N_{\text{C}}}{N_{\text{ref}}}\right)^\lambda}. \tag{3.5.1-40}$$

The exponent factor β is usually 2 for electrons and 1 for holes. For III-V materials, the standard equation for the field-dependent velocity is

$$v_{\mathrm{n}}(|E|) = \frac{\mu_{0\mathrm{D}\Gamma}|E| + v_{\mathrm{sL}}\left(\frac{|E|}{E_0}\right)^{\beta}}{1 + \left(\frac{|E|}{E_0}\right)^{\beta}}, \tag{3.5.1-41}$$

with the critical field E_0, the low-field mobility $\mu_{0\mathrm{D}\Gamma}$ and the saturation velocity in the L-valley v_{sL}. The parameters for (3.5.1-40) and (3.5.1-41) for typical semiconductor materials are listed in tab. 3.2.

For $|E| << E_0$, the velocity is $v_{\mathrm{n}}(|E|) = \mu_{0\mathrm{D}\Gamma}|E|$, corresponding to an ohmic current flow in the Γ valley. Using $E_0 = E_{\mathrm{lim},\Gamma} = v_{\mathrm{s}\Gamma}/\mu_{0\mathrm{D}\Gamma}$ allows to rewrite the velocity function as a function of the normalized electric field $u = |E|/E_{\mathrm{lim},\Gamma}$ as

$$v_{\mathrm{n}}(u) = \frac{v_{\mathrm{s}\Gamma}u + v_{\mathrm{sL}}u^{\beta}}{1 + u^{\beta}}. \tag{3.5.1-42}$$

The relevant variables for (3.5.1-42) are visualized in the schematic view of the velocity-field characteristic in fig. 3.7.

Table 3.2: Parameters for the doping and field dependence of the electron mobility for the standard approximation (3.5.1-41) with (3.5.1-40) as well as (3.5.1-42).

	InP	GaAs	InGaAs	GaN
$\mu_{\mathrm{min}}/(\mathrm{cm}^2\,\mathrm{V}^{-1}\,\mathrm{s}^{-1})$	199	2136	690	176
$\mu_{\mathrm{max}}/(\mathrm{cm}^2\,\mathrm{V}^{-1}\,\mathrm{s}^{-1})$	5172	8467	10018	1409
λ	0.77	0.63	0.91	0.53
$N_{\mathrm{ref}}/(\mathrm{cm}^{-3})$	$2 \cdot 10^{17}$	$7.3 \cdot 10^{17}$	$9 \cdot 10^{17}$	$2.8 \cdot 10^{17}$
$E_0/(\mathrm{kV}\,\mathrm{cm}^{-1})$	14	4	4.09	103.8
$v_{\mathrm{sL}}/(10^7\,\mathrm{cm}\,\mathrm{s}^{-1})$	0.89	0.77	1.26	1
$v_{\mathrm{s}\Gamma}/(10^7\,\mathrm{cm}\,\mathrm{s}^{-1})$	7.241	3.39	4.1	14.62
β	3.63	4	3.85	1.74

Inserting (3.5.1-42) into (3.5.1-37) does unfortunately not allow an analytical solution of the integral. An alternative formulation for the field-dependent velocity is needed. For the integral, it is easier to describe the inverse velocity directly as [SNZ]

$$\frac{1}{v_\mathrm{n}} = \frac{1}{v_{\mathrm{s}\Gamma}} \left[\frac{(1 + u^\beta)^{1/\beta}}{u} + f_\mathrm{v}(u) \right],$$
(3.5.1-43)

where the first term in the bracket corresponds to the $v_\mathrm{n}(E)$ in the Γ-valley, while $f_\mathrm{v}(u)$ is a correction function that takes into account the impact of the second valley and therefore the NDM effect as

$$f_\mathrm{v}(u) = f_\mathrm{v0} \frac{a_\mathrm{fv}(u - u_\mathrm{p})}{\sqrt{1 + [a_\mathrm{fv}(u - u_\mathrm{p})]^2}}.$$
(3.5.1-44)

a_fv and u_p are fit parameters that can be set to fixed values for a given material. f_v0 is related to the ratio of the Γ- and L-valley saturation velocities as

$$f_\mathrm{v0} = \frac{v_{\mathrm{s}\Gamma}}{v_{\mathrm{s}L}} - 1.$$
(3.5.1-45)

The exponential parameter β is not used for fitting, but instead chosen as a fixed value, since closed-form analytical expressions for (3.5.1-37) can only be obtained for $\beta = 1$ and $\beta = 2$. In this work, $\beta = 1$ is chosen due to the relative simplicity of the obtained equations. The choice of β impacts the values of a_fv and u_p via nonlinear fitting. Excellent agreement with the standard equation (3.5.1-41) is obtained, as is demonstrated in fig. 3.8 from a comparison of different modeling approaches and in fig. 3.9 for different materials and doping concentrations.

Fig. 3.8 demonstrates very well that the components of the inverse velocity model do not directly correspond to the Γ- and L-valley velocities; from an analysis of the different contributions to the overall velocity, it becomes clear that the $f_\mathrm{v}(u)$ can become negative and that $v_{\mathrm{s}\Gamma}$ can be smaller than the peak velocity.

As an example, the fitting values obtained for the inverse velocity model for common III-V materials at low doping densities are given in tab. 3.3; these parameters were used in fig. 3.9 for $N_\mathrm{C} = 10^{16}\mathrm{cm}^{-3}$. Note that the Γ-valley saturation velocity in the table is a fitting parameter and does not necessarily represent a physically accurate value. Similarly, E_lim is not equal to E_0; therefore, $v_{\mathrm{s}\Gamma}$ cannot be calculated from $v_{\mathrm{s}\Gamma} = E_0 \mu_{0\mathrm{D}\Gamma}$.

Despite the limited physical reference of the inverse velocity model, though, fig. 3.8 and fig. 3.9 demonstrate that it is possible to use it for modeling a wide range of semiconductor materials and doping values. Unlike a more physics-based formulation, it lends itself well to a closed-form analytical solution of the transit time integral, as will be shown next.

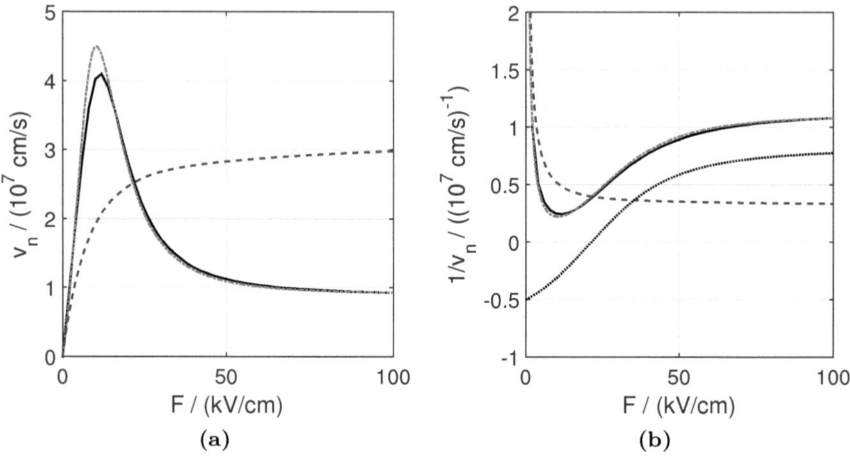

(a)   (b)

Figure 3.8: (a) Electron velocity and (b) inverse electron velocity dependence on the electric field for multiple model equations. Solid line: standard model according to (3.5.1-41). Dash-dotted line: inverse velocity model according to (3.5.1-41). Dashed line: Single-valley contribution (i.e. f_{v0} set to 0). Dotted line in (b): Contribution of second valley term (3.5.1-44).

3.5.1.2 Transit time solution

Regardless of the employed $v_n(E)$ dependence, the integral (3.5.1-37) can be rewritten with the electric field as integration variable, resulting in

$$\tau_{\mathrm{BCv}} = \int_{|E(0)|}^{|E(w_{\mathrm{BC}})|} \left(1 - \frac{x}{w_{\mathrm{BC}}}\right) \frac{1}{v_n(|E|)} \frac{dx}{d|E|} d|E|. \tag{3.5.1-46}$$

Due to the weight factor $(1 - x/w_{\mathrm{BC}})$, the velocity and the corresponding electric field at the base end of the SCR have by far the biggest impact on the integral value. This has been verified by a numerical evaluation of the integral using the electric field profile in collectors with constant doping concentration given by the linear relationship

$$|E(x)| = (E_{\mathrm{jC}} - E_{\mathrm{wc}})(1 - \frac{x}{w_{\mathrm{BC}}}) + E_{\mathrm{wc}}, \tag{3.5.1-47}$$

with the bias-dependent electric field at the BC junction E_{jC} (which is also the peak electric field in the collector at low current densities) and E_{wc}, the field at the end of the collector SCR, which is only different from zero for full depletion, i.e. when the BC SCR reaches the buried layer resp. subcollector and

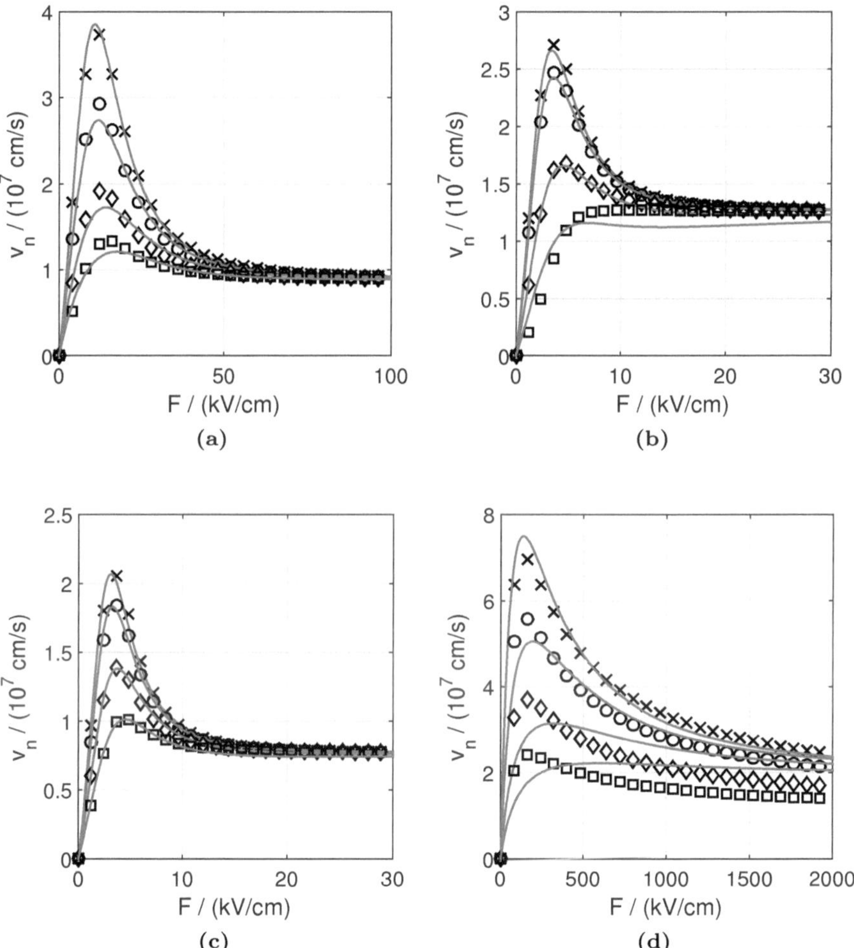

Figure 3.9: Comparison of standard equation (3.5.1-41) (symbols) with simplified inverse velocity model (3.5.1-43) (lines) for multiple materials and dopings. (a) InP, (b) InGaAs lattice-matched to InP, (c) GaAs and (d) GaN. Doping concentrations $N_C = [10^{16}, 10^{17}, 10^{18}, 10^{19}]$ cm^{-3}. Deviations in (d) for high doping values are due to forcing $\beta = 1$, but generally not relevant since collector doping values in III-V HBTs are usually $< 10^{17}$ cm^{-3}.

Table 3.3: Parameters for the field dependence of the electron velocity for the inverse velocity model according to (3.5.1-43).

	InP	GaAs	InGaAs	GaN
$v_{\mathrm{sL}}/(10^7\,\mathrm{cm\,s^{-1}})$	3.6	24.8	35.8	28.2
$E_{\mathrm{lim}}/(\mathrm{kV\,cm^{-1}})$	9.3	40	44.5	117.8
f_{v0}	2.53	28.61	25.63	7.03
a_{fv}	0.332	5.785	6.296	0.084
u_{p}	2.316	0.091	0.08	0.042

nearly immediately drops to zero. The integration limits are mathematically described as

$$
\begin{aligned}
E_{\mathrm{jC}} &= |E(x=0)| = \frac{2}{w_{\mathrm{C}}}\sqrt{V_{\mathrm{PT}}V_{\mathrm{ci}}} \quad && \text{for } V_{\mathrm{ci}} < V_{\mathrm{PT}}, \\
E_{\mathrm{jC}} &= |E(x=0)| = \frac{V_{\mathrm{PT}} + V_{\mathrm{ci}}}{w_{\mathrm{C}}} && \text{otherwise,}
\end{aligned}
\tag{3.5.1-48}
$$

and

$$
\begin{aligned}
E_{\mathrm{wC}} &= |E(w_{\mathrm{BC}})| = 0 && \text{for } V_{\mathrm{ci}} < V_{\mathrm{PT}}, \\
E_{\mathrm{wC}} &= |E(w_{\mathrm{BC}})| = \frac{V_{\mathrm{ci}} - V_{\mathrm{PT}}}{w_{\mathrm{C}}} && \text{otherwise.}
\end{aligned}
\tag{3.5.1-49}
$$

Here, $V_{\mathrm{PT}} = qN_{\mathrm{C}}w_{\mathrm{C}}^2/(2\varepsilon)$ is the collector punch-through voltage and $V_{\mathrm{ci}} = V_{\mathrm{DCi}} - V_{\mathrm{B'C'}}$ is the effective internal collector voltage, with the built-in voltage of the BC junction V_{DCi}. At low current densities, E_{jC}, E_{wc} and w_{BC} depend only on the voltage V_{ci}. The latter is voltage dependent via

$$
\begin{aligned}
w_{\mathrm{BC}} &= \sqrt{\frac{2\varepsilon}{qN_{\mathrm{C}}}V_{\mathrm{ci}}} \quad && \text{for } V_{\mathrm{ci}} < V_{\mathrm{PT}}, \\
w_{\mathrm{BC}} &= w_{\mathrm{C}} && \text{otherwise.}
\end{aligned}
\tag{3.5.1-50}
$$

In order to further reduce the complexity of the analytical solution of the integral, the spatial dependence (3.5.1-47) needs to be reformulated. Despite the weight function $(1-x/w_{\mathrm{BC}})$ in (3.5.1-46), the impact of the inverse velocity values towards the end of the SCR cannot be neglected entirely. $|E(x)|$ is rewritten as

$$
|E(x)| = E_{\mathrm{jc}}(1 - a_{\mathrm{E}}x),
\tag{3.5.1-51}
$$

with the slope of the electric field

$$a_{\mathrm{E}} = \frac{qN_{\mathrm{C}}}{\varepsilon E_{\mathrm{jc}}} = \frac{E_{\mathrm{jc}} - E_{\mathrm{wc}}}{E_{\mathrm{jc}}w_{\mathrm{BC}}} = \left(1 - \frac{E_{\mathrm{wc}}}{w_{\mathrm{BC}}E_{\mathrm{jc}}}\right).$$
(3.5.1-52)

Evaluating the derivative w.r.t. x in (3.5.1-46) yields

$$\frac{dx}{d|E(x)|} = -E_{\mathrm{jc}}a_{\mathrm{E}} = -\frac{E_{\mathrm{jc}} - E_{\mathrm{wc}}}{w_{\mathrm{BC}}},$$
(3.5.1-53)

and replacing the remaining x-dependent term

$$1 - \frac{x}{w_{\mathrm{BC}}} = 1 - \frac{E_{\mathrm{jc}} - |E(x)|}{E_{\mathrm{jc}} - E_{\mathrm{wc}}} = \frac{|E(x)| - E_{\mathrm{wc}}}{E_{\mathrm{jc}} - E_{\mathrm{wc}}}$$
(3.5.1-54)

yields the integral

$$\tau_{\mathrm{BCv}} = -\frac{w_{\mathrm{BC}}}{E_{\mathrm{jc}} - E_{\mathrm{wc}}} \int_{E_{\mathrm{jc}}}^{E_{\mathrm{wc}}} \frac{|E(x)| - E_{\mathrm{wc}}}{E_{\mathrm{jc}} - E_{\mathrm{wc}}} \frac{1}{v(|E|)} d|E|.$$
(3.5.1-55)

Introducing the normalized variables $u_{\mathrm{jc}} = E_{\mathrm{jc}}/E_{\mathrm{lim}}$ and $u_{\mathrm{wc}} = E_{\mathrm{wc}}/E_{\mathrm{lim}}$ allows to write the normalized form of (3.5.1-55) as

$$\tau_{\mathrm{BCv}} = -\frac{w_{\mathrm{BC}}}{u_{\mathrm{jc}} - u_{\mathrm{wc}}} \int_{u_{\mathrm{jc}}}^{u_{\mathrm{wc}}} \frac{u - u_{\mathrm{wc}}}{u_{\mathrm{jc}} - u_{\mathrm{wc}}} \frac{1}{v(u)} du.$$
(3.5.1-56)

Since the inverse velocity model in (3.5.1-43) has been expressed as the sum of two terms, the evaluation of the integral can be split into two cases; the first case uses just the single-valley expression, valid for the low electric fields in III-V materials and for all fields for SiGe. The second case adds the $f_{\mathrm{v}}(u)$ related term as a correction function for the two-valley velocity. The single-valley equation is derived here first.

Inserting (3.5.1-43) into (3.5.1-56), neglecting the f_{v}-term for now, yields

$$\tau_{\mathrm{BCv}} = -\frac{w_{\mathrm{BC}}}{v_{\mathrm{s\Gamma}}(u_{\mathrm{jc}} - u_{\mathrm{wc}})^2} \int_{u_{\mathrm{jc}}}^{u_{\mathrm{wc}}} \frac{u - u_{\mathrm{wc}}}{u} (1 + u^\beta)^{1/\beta} du.$$
(3.5.1-57)

Keeping in mind that $\beta = 1$, the closed-form solution reads

$$\tau_{\mathrm{BCv}} = \frac{w_{\mathrm{BC}}}{2v_{\mathrm{s\Gamma}}} \left[1 + 2\frac{u_{\mathrm{jc}} - u_{\mathrm{wc}} + u_{\mathrm{wc}} \ln\left(\frac{u_{\mathrm{wc}}}{u_{\mathrm{jc}}}\right)}{(u_{\mathrm{jc}} - u_{\mathrm{wc}})^2} \right].$$
(3.5.1-58)

For the asymptotic case of high voltages and corresponding fields (i.e. $u_{\mathrm{jc}} \gg 1$ and $u_{\mathrm{jc}} \gg u_{\mathrm{wc}}$), the classical result

$$\tau_{\mathrm{BCv}} \approx \frac{w_{\mathrm{BC}}}{2v_{\mathrm{s\Gamma}}}$$
(3.5.1-59)

is obtained. Note that for very large voltages, $E_{jc} = E_{wc} + w_{BC}\frac{qN_C}{\varepsilon}$, i.e. the electric field at the junction has a constant offset from the field at the end of the SCR determined by the collector doping density and width. For all practical voltages and transistors, though, $u_{jc} \gg u_{wc}$ holds.

For the case of low voltages (i.e. $u_{wc} = 0$), one obtains

$$\tau_{BCv} \approx \frac{w_{BC}}{2v_{s\Gamma}}\left[1 + \frac{1}{u_{jc}}\right] \approx \frac{w_{BC}}{2v_{s\Gamma}}\frac{E_{lim}}{E_{jc}}. \tag{3.5.1-60}$$

Evaluating the second term in (3.5.1-43) leads to a voltage-dependent change in the total transit time as

$$\Delta\tau_{BCv} = -\frac{w_{BC}}{v_{s\Gamma}(u_{jc} - u_{wc})^2}\int_{u_{jc}}^{u_{wc}}(u - u_{wc})f_v(u)du. \tag{3.5.1-61}$$

The closed form analytical solution is

$$\Delta\tau_{BCv} = -\frac{w_{BC}}{2v_{s\Gamma}(u_{jcp} - u_{wcp})^2}f_{v0}\cdot\left[u_{wcp}\sqrt{1 + u_{wcp}^2}+\right.$$
$$\left.(u_{jcp} - 2u_{wcp})\sqrt{1 + u_{jcp}^2} - \ln\left(\frac{u_{jcp} + \sqrt{1 + u_{jcp}^2}}{u_{wcp} + \sqrt{1 + u_{wcp}^2}}\right)\right], \tag{3.5.1-62}$$

with the auxiliary variables

$$u_{jcp} = a_{fv}(u_{jc} - u_p) \text{ and } u_{wcp} = a_{fv}(u_{wc} - u_p). \tag{3.5.1-63}$$

Generally, the transit time is increased, though for $u < u_p$ the opposite is true (c.f. fig. 3.8b). For the limit of very high fields, (3.5.1-62) results in

$$\Delta\tau_{BCv} \approx \frac{w_{BC}}{2v_{s\Gamma}}f_{v0}, \tag{3.5.1-64}$$

while for very low fields, one obtains

$$\Delta\tau_{BCv} \approx \frac{w_{BC}a_{fv}}{2v_{s\Gamma}u_{jc}}\frac{f_{v0}}{\sqrt{1 + (a_{fv}u_p)^2}}. \tag{3.5.1-65}$$

The total collector transit time through the BC SCR consists of the sum of (3.5.1-58) and (3.5.1-62)

$$\tau_{BC} = \tau_{BCv} + \Delta\tau_{BCv}. \tag{3.5.1-66}$$

For very high fields in III-V materials, one obtains for the total transit time

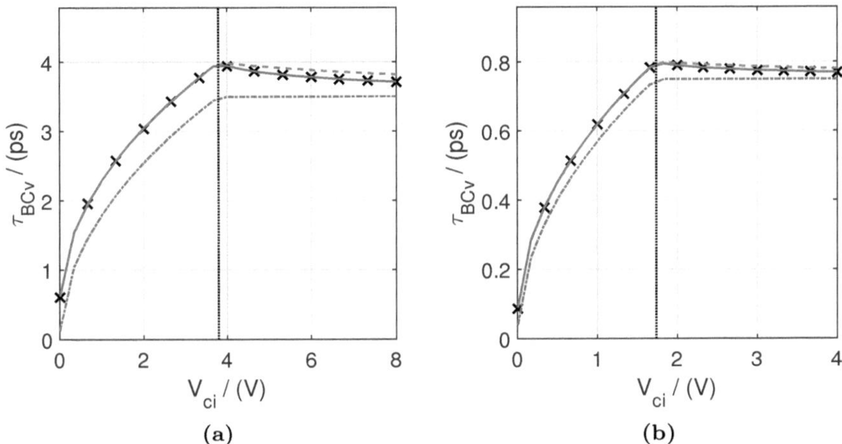

Figure 3.10: Comparison of the voltage dependent collector transit time in silicon calculated from the numerical integration of (3.5.1-57) (crosses) with the analytical solution (3.5.1-58) (solid line), an evaluation of (3.5.1-58) with $E_{\mathrm{wc}} = 0\,\mathrm{V\,cm}^{-1}$ (dashed line), and the classical result (3.5.1-59) dash-dotted line) for different collector parameters: (a) $N_{\mathrm{C}} = 10^{16}\mathrm{cm}^{-3}$ and $w_{\mathrm{C}} = 0.7\,\mu\mathrm{m}$; (b) $N_{\mathrm{C}} = 10^{17}\ \mathrm{cm}^{-3}$ and $w_{\mathrm{C}} = 0.15\,\mu\mathrm{m}$. In both cases, the collector material is Si with $f_{\mathrm{v0}} = 0$, i.e. no NDM effect. The vertical line marks the punch-through voltage.

$$\tau_{\mathrm{BC}} \approx \frac{w_{\mathrm{BC}}}{2v_{\mathrm{s\Gamma}}}f_{\mathrm{v0}} + \frac{w_{\mathrm{BC}}}{2v_{\mathrm{s\Gamma}}} = \frac{w_{\mathrm{BC}}}{2v_{\mathrm{s\Gamma}}}(1 + f_{\mathrm{v0}}) = \frac{w_{\mathrm{BC}}}{2v_{\mathrm{sL}}}, \tag{3.5.1-67}$$

which corresponds to the classical solution with the saturation velocity of the L-valley.

The analytical solution (3.5.1-66) is first evaluated for silicon ($f_{\mathrm{v0}} = 0$) and then for III-V materials ($f_{\mathrm{v0}} > 0$).

Fig. 3.10 shows the transit time for different parameters of a silicon collector region. The result from (3.5.1-58) is identical with the numerical evaluation of the integral of (3.5.1-57), which verifies the obtained analytical solution. The transit time first increases with V_{ci} due to the increasing SCR width the impact of which is stronger than the also increasing velocity. The peak value is reached for the punch-through case. The voltage independent SCR width together with the still increasing velocity then leads to a more or less slight decrease of τ_{BC}. This is different for the solution from the classical expression (3.5.1-59), which always assumes a constant (e.g. saturation) ve-

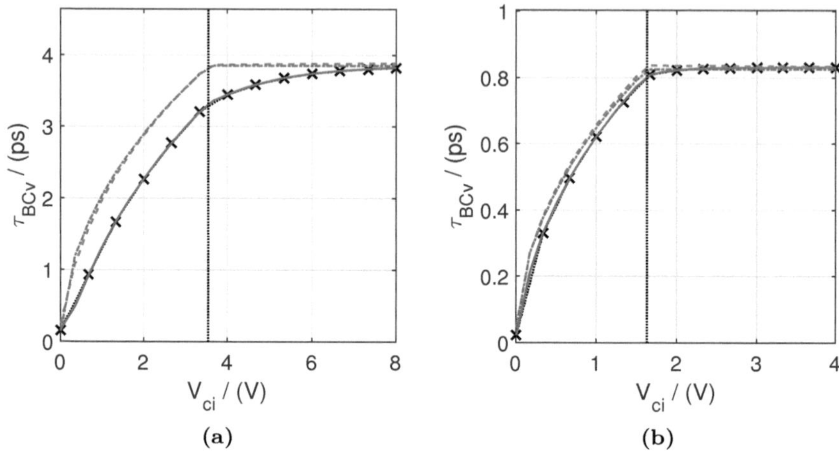

(a)   (b)

Figure 3.11: Comparison of the voltage dependent collector transit time in InP calculated from the numerical integration of (3.5.1-57) (crosses) with the analytical solution (3.5.1-66) (solid line), an evaluation of (3.5.1-66) with $E_{wc} = 0\,\mathrm{V\,cm^{-1}}$ (dashed line), and, the classical result (3.5.1-59) dash-dotted line) for different collector parameters: (a) $N_C = 10^{16}\,\mathrm{cm^{-3}}$ and $w_C = 0.7\,\mu\mathrm{m}$; (b) $N_C = 10^{17}\,\mathrm{cm^{-3}}$ and $w_C = 0.15\,\mu\mathrm{m}$. In both cases, the collector material is InP. The vertical line marks the punch-through voltage.

locity, leading also to a lower τ_{BC} value a low V_{ci}. Simplifying the analytical expression (3.5.1-58) by assuming $E_{wc} = 0\,\mathrm{V\,cm^{-1}}$ is possible with a small error for high voltages. As fig. 3.10 shows, the corresponding curve is on top of the exact solution up to the punch-through voltage, but then starts to deviate.

Fig. 3.11 shows the transit time for an InP collector region with different parameters. Here also the analytical solution (3.5.1-66) is verified by the exact agreement with the numerical evaluation of the corresponding integral in (3.5.1-57) and (3.5.1-61). At low V_{ci} a similar trend as in the silicon case above is observed. However, beyond the punch-through voltage, τ_{BC} keeps increasing due to the field dependent decrease of the velocity. Moreover, the classical expression yields larger τ_{BC} values as long as the lower velocity v_{sL} is inserted. The simplified expression resulting from (3.5.1-57) and (3.5.1-61) when assuming $E_{wc} = 0\,\mathrm{V\,cm^{-1}}$ is nearly identical with the classical solution, but shows a more abrupt transition after punch-through and a different bias dependent slope for the high-voltage collector, but very small deviations for the high-speed case.

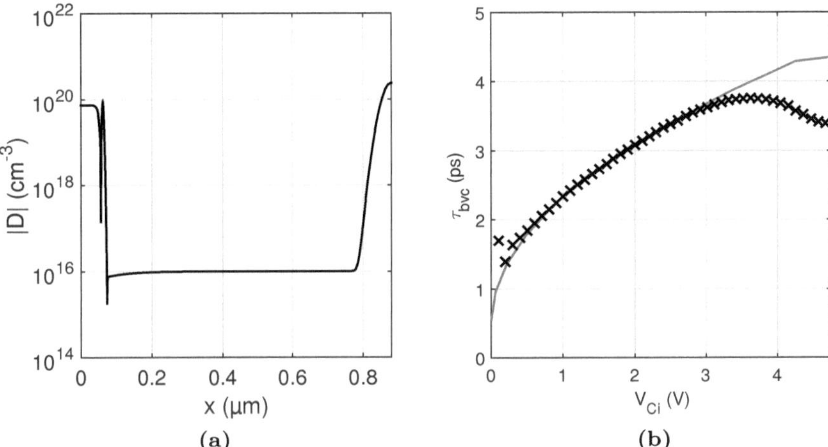

Figure 3.12: SiGe device simulation. (a) Simulated doping profile and (b) resulting transit time from simulation (symbols) and analytical model (line).

For SiGe HBTs, verification of the model by numerical device is possible. The simulated doping profile and resulting transit time, compared with the analytical model, are shown in fig. 3.12. The simulation result and the analytical model agree very well, except for very high voltages. The reason for the deviation is the transition to the buried layer, which was considered to be abrupt in the derivation above; however, in the simulation, the additional space charge region at the transition from high to low doping influences the determined transit time values.

3.5.1.3 Application to experimental data

In order to verify the theoretical calculations in the absence of reliable simulations that show the NDM effect, since BTE simulations do not show good agreement with measurements, the model is applied to measured data. The external parasitic elements are deembedded as described in section 4.4.6. Then, the low-bias transit time is determined from a linear extrapolation of $1/(2\pi f_{\mathrm{Ti}})$ vs. $1/J_{\mathrm{C}}$. The base transit time is assumed to be bias-independent due to the high base doping, estimated from knowledge about the structure and subtracted from the extracted value. The parameters N_{C}, w_{c}, f_{v0} and u_{p} are treated as model parameters and V_{jCi}, the built-in voltage of the internal BC junction capacitance, is used with the applied V_{BC} to calculate V_{ci}. The comparison of the model with the measurement is shown in fig. 3.13; the model parameters are listed in tab. 3.13. They correspond well to the ones derived from theory except the collector doping, because the collector doping spikes require insertion of an *effective* doping value. Overall, excellent agreement is obtained, though for positive V_{BC}, the assumption of a constant base transit time for the InP device breaks down.

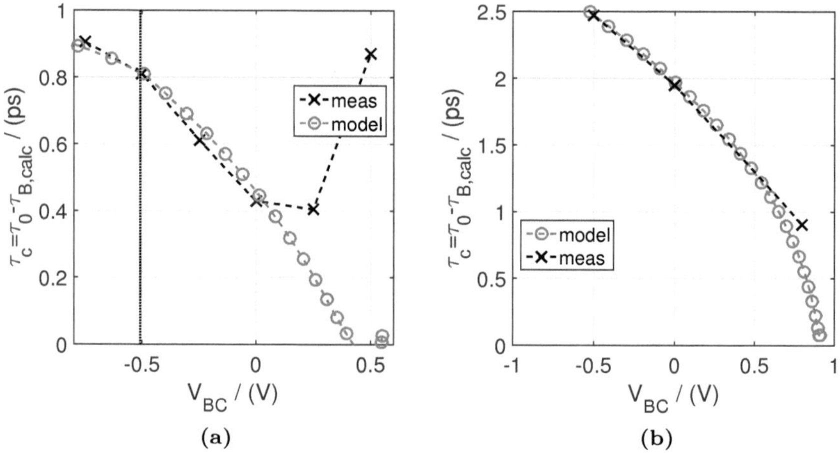

(a)   (b)

Figure 3.13: Comparison of the derived model with parameters according to tab. 3.4 to measurements for (a) a high-speed InP device and (b) a high-voltage GaAs device.

Table 3.4: Model parameters used for fig. 3.13.

HBT type	$w_c/\mu m$	N_C/cm^3	V_{jCi}/V	f_{v0}	u_{p0}
InP	0.15	6.5e16	0.55	3.5	5
GaAs	0.6	1.3e16	0.91	28.6	0.09

3.5.1.4 Compact model implementation

The equations obtained so far are implemented in the compact model using the new physics-based parameters $\tau_{\mathrm{BCv0}} = \frac{w_{\mathrm{c}}}{2v_{\mathrm{s\Gamma}}}$, $V_0 = E_{\mathrm{lim}}w_{\mathrm{c}}$, f_{v0}, a_{fv} and u_{p}. The latter two can likely be set to constant values, reducing the total number of new model parameters to three.

The (Γ valley) collector transit time τ_{BCv} is modeled as

$$\tau_{\mathrm{BCv}} = \frac{\tau_{\mathrm{BCv0}}}{c_{\mathrm{PT}}}\left(1 + \frac{V_0}{2c_{\mathrm{PT}}V_{\mathrm{ci}}}\right) \qquad \text{for } V_{\mathrm{ci}} \leq V_{\mathrm{PT}} \text{ and}$$

$$\tau_{\mathrm{BCv}} = \tau_{\mathrm{BCv0}}\left(1 + \frac{\frac{2V_{\mathrm{PT}}}{V_0} + \frac{V_{\mathrm{ci}}-V_{\mathrm{PT}}}{V_0}\cdot\ln\left(\frac{V_{\mathrm{ci}}+V_{\mathrm{PT}}}{V_{\mathrm{ci}}-V_{\mathrm{PT}}}\right)}{(2V_{\mathrm{PT}}/V_0)^2}\right) \qquad \text{otherwise.}$$

$$(3.5.1\text{-}68)$$

The offset caused by the NDM effect is described according to (3.5.1-62) with the components modeled as

$$u_{\mathrm{wcp}} = -u_{\mathrm{p}}a_{\mathrm{afv}} \qquad \text{for } V_{\mathrm{ci}} < V_{\mathrm{PT}} \text{ and}$$

$$u_{\mathrm{wcp}} = \left(\frac{V_{\mathrm{ci}} - V_{\mathrm{PT}}}{V_0} - u_{\mathrm{p}}\right) \qquad \text{otherwise,}$$

$$(3.5.1\text{-}69)$$

$$u_{\mathrm{jcp}} = a_{\mathrm{fv}}\left(\frac{2V_{\mathrm{ci}}}{V_0}c_{\mathrm{PT}} - u_{\mathrm{p}}\right) \qquad \text{for } V_{\mathrm{ci}} < V_{\mathrm{PT}} \text{ and}$$

$$u_{\mathrm{jcp}} = a_{\mathrm{fv}}\left(\frac{V_{\mathrm{ci}} + V_{\mathrm{PT}}}{V_0} - u_{\mathrm{p}}\right) \qquad \text{otherwise,}$$

$$(3.5.1\text{-}70)$$

and summarizing the first factor in the equation as

$$\Delta\tau_{\mathrm{BCv0}} = \frac{\tau_{\mathrm{BCv0}}}{c_{\mathrm{PT}}}f_{\mathrm{v0}} \qquad \text{for } V_{\mathrm{ci}} < V_{\mathrm{PT}} \text{ and}$$

$$\Delta\tau_{\mathrm{BCv0}} = \tau_{\mathrm{BCv0}}c_{\mathrm{PT}}f_{\mathrm{v0}} \qquad \text{otherwise.}$$

$$(3.5.1\text{-}71)$$

In those model equations, the normalized punch-through capacitance

$$c_{\mathrm{PT}} = \frac{C_{\mathrm{jC}}}{C_{\mathrm{jCPT}}} = \frac{C_{\mathrm{jC}}}{C_{\mathrm{jC}}(V_{\mathrm{ci}} = V_{\mathrm{PT}})} \qquad (3.5.1\text{-}72)$$

was used to describe the collector SCR width. Note, that this is a preliminary implementation used in this work in order to evaluate the model in chapter 5 and may not be numerically stable under all conditions.

3.5.2 Medium current transit time

Having obtained a formulation for the low-bias collector transit time, a current dependence is still required for a full model of the NDM effect. Only the high-voltage case is considered here, since, in the low voltage region (i.e. $E(J_{\mathrm{lim}}) < E_{\mathrm{p}}$, with the characteristic current density J_{lim} for which the slope of the electric field in the collector is zero), the same considerations as for silicon apply (e.g. [SC10]).

As a preliminary consideration, the influence of the current on the electric field in the collector is shown in fig. 3.14 for four different current densities. In case (1), J_{C} is low and has little influence on the doping-dependent space charge; the electric field is high in the entire collector. For case (2), $J_{\mathrm{C}} = J_{\mathrm{lim}}$, i.e. the electrons carrying the collector current, traveling with L-valley saturation velocity, compensate the space charge and the electric field is constant. Case (3) shows the onset of the NDM effect; the electric field has an inverted slope, i.e. the peak field is located at the collector end of the SCR, but is sufficiently strong to carry the current at L-valley saturation velocity. In SiGe, without NDM, this indicates the onset of the Kirk effect if the electric field if E_{jC} equals E_{lim}, since the carrier velocity decreases at the BC junction with a further current increase. For III-V materials, however, the carrier velocity *increases* before is begins to drop. This is shown in case (4), where the electric field has dropped below E_{sat} at the BC junction. Since the current is constant, this means that the carrier density and therefore the space charge $\rho = q \left(N_{\mathrm{C}} - \frac{J_{\mathrm{C}}}{qv_{\mathrm{n}}} \right)$ decrease, causing a lower slope of the electric field. According to (3.5.1-37), the transit time decreases due to the increased carrier velocity. Case (5) represents the onset of high current effects in III-V materials; E_{jC} has decreased so much that the carriers travel at peak velocity and any further decrease leads to a lower velocity. The high current region in III-V materials follows the same considerations as silicon.

For non-zero current and a field-dependent velocity (i.e. the carriers do not move at saturation velocity; for SiGe, this means $E(x) < E_{\mathrm{lim}}$ while for III-V materials, it means that $E(x) < E_{\mathrm{sat}}.$), the spatial distribution of the electric field depends on the solution of the differential equation

$$\frac{dE}{dx} = q(N_{\mathrm{C}} - \frac{J_{\mathrm{C}}}{qv_{\mathrm{n}}(E)}).\tag{3.5.2-73}$$

However, while (3.5.2-73) is solvable for multiple approaches for modeling the field-dependent velocity, the resulting integral for the transit time (3.5.1-56) generally has no analytical solutions for non-constant velocities [SC10]. Therefore, simplified solutions must be considered. Multiple approaches are discussed in this section.

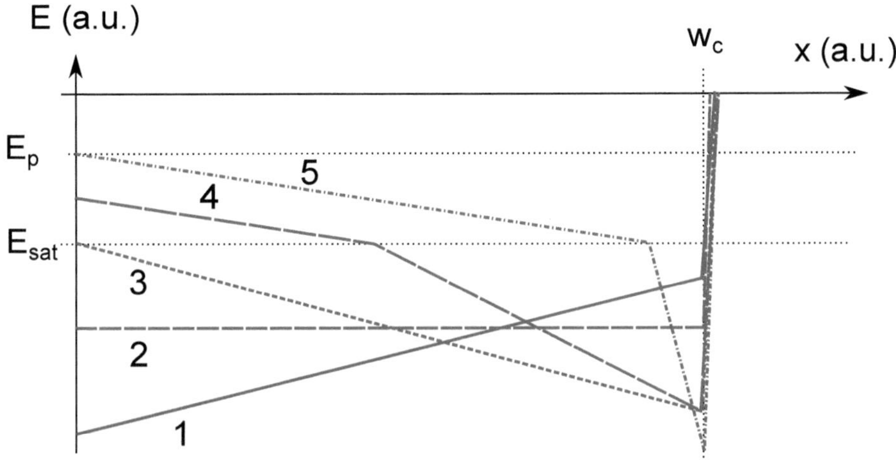

Figure 3.14: Schematic representation of the electric field in the collector for different current densities: (1) low current, no impact on the space charge; (2) $J_C = J_{lim}$; (3) $J_C = J_{NDM}$, onset of current-dependent NDM effect; (4) medium current density; (5) $J_C = J_{CK}$.

Since no device simulations that take the NDM effect into account are available, the verification of the models developed here depends on measured data. For this purpose, an InP and a GaAs process are used. In both cases, the transit frequency exhibits the cross-over in the medium bias range that is typical for the NDM effect, as shown in fig. 3.15.

From the small signal data, the parasitic and external elements are deembedded as described in section 4.4.6 to obtain the internal transit frequency $f_{t,int}$. The extrapolation of the internal transit time $\tau_f = 1/(2\pi f_{t,int})$ from the low-bias region w.r.t. $1/J_C$ and the resulting $\Delta\tau_{BCv}$ are shown in fig. 3.16. The collector transit time is defined here as the difference between linear extrapolation and the measured transit time between low current and the minimum of τ_f.

3.5.2.1 Empirical model

A simple empirical model can be implemented by noting the following trends of the measured data:

- The current dependent collector transit time is V_{BC}-dependent, with a maximum value $\Delta\tau_{BCv0}$ and a minimum of zero. The lower V_{BC}, the higher the transit time.

- With increasing current, the transit time decreases to zero.

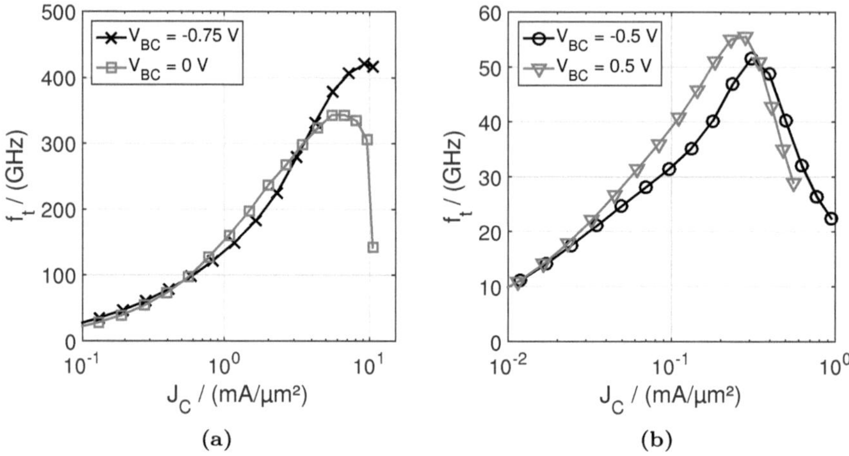

(a)   (b)

Figure 3.15: Transit frequency for high and low reverse BC bias, demonstrating the cross-over (i.e. lower f_T for higher negative V_BC at medium current densities) for (a) InP and (b) GaAs technology.

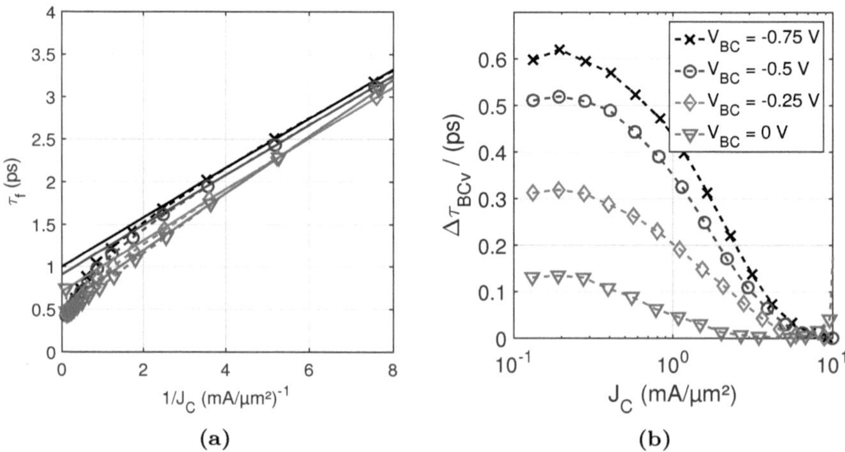

(a)   (b)

Figure 3.16: (a) Measured internal transit time (symbols) and linear extrapolation (lines) and (b) resulting $\Delta\tau_\mathrm{BCv}$ as the difference between extrapolation and measured transit time for the same InP process as in 3.15a . The legend in (b) is valid for both figures.

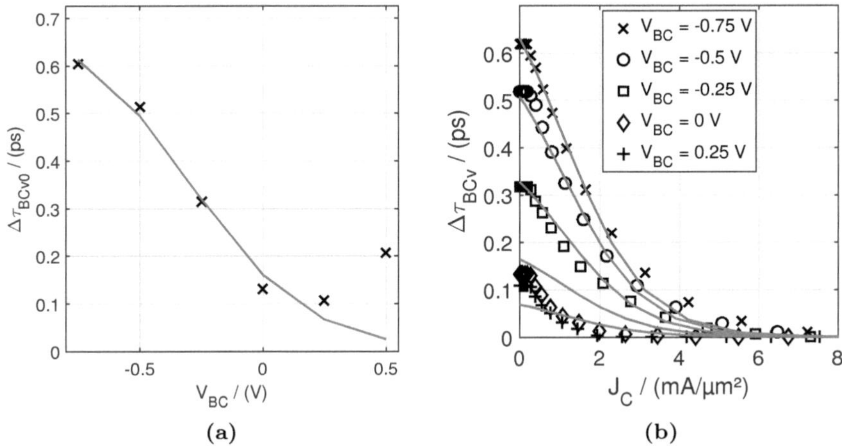

Figure 3.17: (a) V_{BC} dependence of $\Delta\tau_{BCv0} = max(\Delta\tau_{BCv})$ at $J_C = 0$ and (b) current dependence at various V_{BC}. Symbols: measurement of InP process and lines: model according to (3.5.2-75).

Furthermore, a description for the charge is needed, i.e. the transit time description needs to be integrable. From the capacitance model described in section 3.4, transition functions that fulfill this requirement are already known. Implementing them, the full model reads

$$\Delta\tau_{BCv} = \Delta\tau_{BCv0} \cdot \frac{\exp\left(\frac{V_{NDM}-V_{BC}}{A_{NDM,V}}\right)}{1 + \exp\left(\frac{V_{NDM}-V_{BC}}{A_{NDM,V}}\right)}, \tag{3.5.2-74}$$

$$Q_{NDM} = \Delta\tau_{BCv} \cdot J_{C,\text{lim}} \tag{3.5.2-75}$$

and

$$J_{C,\text{lim}} = J_{NDM} - A_{NDM,I} \ln\left[1 + \exp\left(\frac{J_{NDM} - J_C}{A_{NDM,I}}\right)\right], \tag{3.5.2-76}$$

with the model parameters V_{NDM}, J_{NDM}, $A_{NDM,I}$, $A_{NDM,V}$ and $\Delta\tau_{BCv0}$. If required, J_{NDM} can be made voltage dependent using the additional model parameters k_{NDM} and using $J_{NDM} = J_{NDM0} + k_{NDM}V_{BC}$.

The model is simple and efficient, with intuitive parameters and it is easy to implement in most compact models. Its sole disadvantage is that it is purely empirical, lacking all physical basis and initial parameter values based on structure and material data. An application example is shown in fig. 3.17.

3.5.2.2 Semi-empirical model

For an improved physical basis of the model, the equations developed for $\Delta\tau_{\mathrm{BCv}}$ in the section 3.5.1.4 can be used instead of the simple empirical model (3.5.2-74). Note that only the $\Delta\tau_{\mathrm{BCv}}$ term is current dependent. This introduces a number of additional parameters and makes the bias dependence more complicated. However, physics-based parameters, for which reasonable values are known from structure data, are generally worth the trade-off. This version of the model has been implemented for the comparisons shown in chapter 5.

3.5.2.3 Linear current-field dependence

In keeping with the assumption of a trapezoidal field in the collector, i.e. a constant space charge, which is identical to assuming a constant carrier velocity, it can be assumed for the description of $E(x, J_{\mathrm{C}})$ that all carriers move at $v_{\mathrm{n}} = v_{\mathrm{sat,L}}$. Furthermore, only the case of high collector voltages is considered here, since the case of low voltages is covered by the description of the high current effects as for SiGe devices in HICUM. Then, the relevant field values can be calculated as

$$u_{\mathrm{jc}} = -\frac{V_{\mathrm{PT}}}{J_{\mathrm{lim}}V_0}J_{\mathrm{C}} + \frac{V_{\mathrm{ci}} + V_{\mathrm{PT}}}{V_0} \tag{3.5.2-77}$$

and

$$u_{\mathrm{wc}} = \frac{V_{\mathrm{PT}}}{J_{\mathrm{lim}}V_0}J_{\mathrm{C}} + \frac{V_{\mathrm{ci}} - V_{\mathrm{PT}}}{V_0}. \tag{3.5.2-78}$$

with the model parameter $V_0 = E_{\mathrm{lim}}w_{\mathrm{c}}$.

While the integral of (3.5.1-56) w.r.t. J_{C} with (3.5.2-77) and (3.5.2-78) can be solved, even with this very simple relationship between the field and the current, the determination of the collector charge becomes too complicated for a compact model. A more simple relationship for the velocity than (3.5.1-43) is needed.

Since the field in the collector is assumed to be larger than E_{p}, the field at peak velocity, a linear inverse velocity interpolation

$$\frac{1}{v_{\mathrm{n}}} = \frac{1}{v_{\mathrm{p}}} + \frac{1/v_{\mathrm{s}} - 1/v_{\mathrm{p}}}{u_{\mathrm{s}} - u_{\mathrm{p}}}(u - u_{\mathrm{p}}) \tag{3.5.2-79}$$

is used, with the peak and saturation velocities v_{p} and v_{s} and the normalized fields at which they occur u_{p} and u_{s}, respectively. Then, using the same integration steps as sketched out in section 3.5.1.2, the resulting transit time is

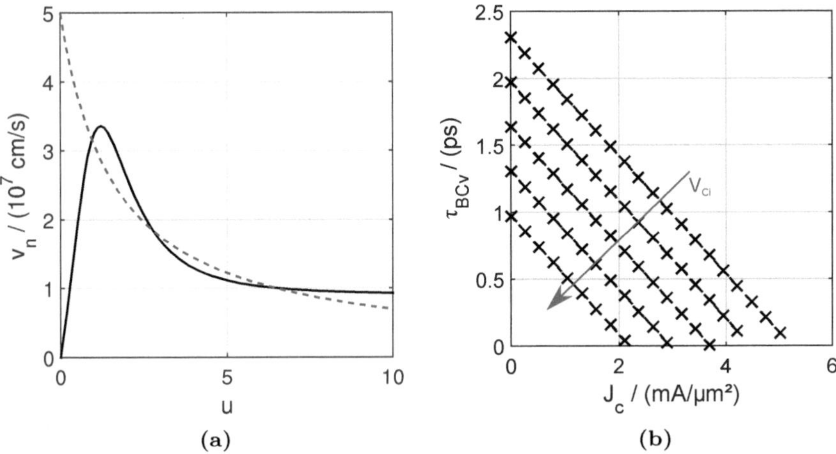

Figure 3.18: (a) Carrier velocity as a function of the normalized electric field. Solid line: velocity according to (3.5.1-41) and dashed line: velocity according to (3.5.2-79). (b) Resulting collector transit time as a function of the collector current density for $V_{\mathrm{Ci}} = 0.85\ldots3.35$ V.

Table 3.5: Interpolation points for linear inverse velocity as shown in fig. 3.18a and structure parameters used for fig. 3.18b.

w_c	N_C	v_p	v_s	E_p	E_s
$/\mu\mathrm{m}$	$/(\mathrm{cm}^3)$	$/(\mathrm{cm\,s}^{-1})$	$/(\mathrm{cm\,s}^{-1})$	$/(\mathrm{kV\,cm}^{-1})$	$/(\mathrm{kV\,cm}^{-1})$
0.2	$3\cdot10^{16}$	$3\cdot10^7$	$1\cdot10^7$	10	60

$$\tau_{\mathrm{BCv}} = \frac{w_\mathrm{c}}{2v_\mathrm{p}} + \frac{1}{6}\frac{w_\mathrm{c}(1/v_\mathrm{s} - 1/v_\mathrm{p})}{u_\mathrm{s} - u_\mathrm{p}}(2u_\mathrm{jc} - 3u_\mathrm{p} + u_\mathrm{wc}), \qquad (3.5.2\text{-}80)$$

which can easily be integrated to obtain the collector charge. However, the combination of an assumed linear electric field in the collector and the linear field dependence of the velocity lead to a linear transit time-current relation, which does not agree well with the values obtained from measurements. The model is therefore not compared to the measurements directly.

An example for the comparison of the linear inverse velocity formulation with the accurate velocity according to (3.5.1-41) and the resulting transit time is shown in fig. 3.18

3.5.2.4 Step function model

In order to obtain a physics-based solution for (3.5.2-73), it is assumed that
the carrier velocity is

$$
\begin{aligned}
v_\mathrm{n} &= v_\mathrm{sat} && \text{for } E \geq E_\mathrm{sat} \\
v_\mathrm{n} &= v_\mathrm{p} && \text{for } E_\mathrm{p} \geq E < E_\mathrm{sat} \\
v_\mathrm{n} &= (v_\mathrm{p}/E_\mathrm{p}) \cdot E && \text{for } E < E_\mathrm{p},
\end{aligned}
\tag{3.5.2-81}
$$

as visualized in fig. 3.19. Since $E_\mathrm{jC} = E_\mathrm{p}$ represents the onset of high current
effects, $E < E_\mathrm{p}$ is not considered here.

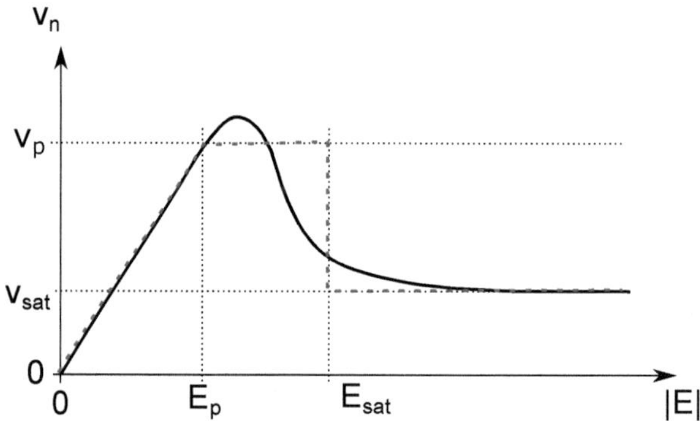

Figure 3.19: Schematic view of field-dependent velocity (solid line) and
velocity step function (dashed line) assumed for the transit time derivation.

The collector is thus divided into a region $x < x_\mathrm{t}$, in which the carriers
travel at v_p, and a region $x \geq x_\mathrm{t}$, in which carriers travel at v_sat. The collector
transit time integral can then be split in two separate integrals dependent on
the transition point x_t as

$$
\begin{aligned}
\tau_\mathrm{BCv} &= \int_0^{w_\mathrm{c}} \left(1 - \frac{x}{w_\mathrm{c}}\right) \frac{1}{v_\mathrm{n}(x)}\,\mathrm{d}x \\
&= \int_0^{x_\mathrm{t}} \left(1 - \frac{x}{w_\mathrm{c}}\right) \frac{1}{v_\mathrm{p}}\,\mathrm{d}x + \int_{x_\mathrm{t}}^{w_\mathrm{c}} \left(1 - \frac{x}{w_\mathrm{c}}\right) \frac{1}{v_\mathrm{s}}\,\mathrm{d}x \\
&= \frac{1}{2w_\mathrm{c}} \left[\frac{1}{v_\mathrm{p}}(2w_\mathrm{c}x_\mathrm{t} - x_\mathrm{t}^2) + \frac{1}{v_\mathrm{sat}}(w_\mathrm{c} - x_\mathrm{t}^2)\right].
\end{aligned}
\tag{3.5.2-82}
$$

Next, the current dependence of the transition point needs to be determined. By definition, the electric field at x_t is

$$
E(x_\mathrm{t}) = E_\mathrm{sat}.
\tag{3.5.2-83}
$$

The slope of the electric field is determined by the space charge, which in turn is dependent on the electron velocity, resulting in

$$E(x) = E_{jC} + \int_0^x \frac{q}{\varepsilon}\left(N_C - \frac{J_C}{qv_p}\right)dx' = E_{jC} + x\frac{q}{\varepsilon}\left(-\frac{J_C}{qv_p} + N_C\right) \quad (3.5.2\text{-}84)$$

for $x < x_t$ and

$$E(x) = E(x_t) + \int_{x_t}^x \frac{q}{\varepsilon}\left(N_C - \frac{J_C}{qv_{sat}}\right)dx' = E_{sat} + (x - x_t)\frac{q}{\varepsilon}\left(-\frac{J_C}{qv_{sat}} + N_C\right)$$
$$(3.5.2\text{-}85)$$

otherwise, using the shorthand $\varepsilon = \varepsilon_0 \varepsilon_r$, the collector current density J_C, which is approximately equal to the transfer current under forward active conditions and the parameter as defined in fig. 3.5.2-82.

Furthermore, the integral over the electric field must be equal to V_{Ci}. Therefore,

$$\begin{aligned}
V_{Ci} &= -\int_0^{w_c} E(x)dx \\
&= E_{jC}w_c + w_c^2 \frac{q}{2\varepsilon}\left(\frac{J_C}{v_{sat} - N_C}\right) \\
&\quad + \frac{x_t^2 q}{2\varepsilon}\left(\frac{J_C}{v_{sat}} - \frac{J_C}{v_p}\right) + \frac{x_t w_c q}{\varepsilon}\left(\frac{J_C}{v_{sat}} - \frac{J_C}{v_p}\right)
\end{aligned} \quad (3.5.2\text{-}86)$$

The electric field at the BC junction can be calculated from (3.5.2-84) as

$$E_{jC} = E_{sat} - \frac{x_t}{\varepsilon}\left(\frac{J_C}{v_p} - qN_c\right). \quad (3.5.2\text{-}87)$$

Inserting (3.5.2-87) into (3.5.2-86) and solving for x_t yields

$$x_t = v_p w_c \frac{J_C - qv_{sat}N_C}{J_C(v_p - v_{sat})} - $$
$$\sqrt{\left[\frac{(v_p w_c)(J_C - qv_{sat}N_C)}{J_C(v_p - v_{sat})}\right]^2 + \frac{v_p v_{sat}\left[2\varepsilon V_{Ci} - 2\varepsilon w_c E_{sat} - w_c^2(J_C/v_{sat} - qN_C)\right]}{J_C(v_p - v_{sat})}}$$
$$(3.5.2\text{-}88)$$

(3.5.2-88) allows solving (3.5.2-82) depending only on physical and technological parameters and terminal quantities. However, (3.5.2-88) is very complicated and not suitable for compact model implementation. By defining the physics-based model parameters

$$k = \frac{v_{\mathrm{p}}}{v_{\mathrm{sat}}}$$

$$\tau_{\mathrm{c0}} = \frac{w_{\mathrm{c}}}{2v_{\mathrm{sat}}}$$

$$J_{\mathrm{lim}} = q v_{\mathrm{sat}} N_{\mathrm{C}} \tag{3.5.2-89}$$

$$G_{\mathrm{vci}} = \frac{\varepsilon}{\tau_{\mathrm{c0}} w_{\mathrm{c}}} = \frac{J_{\mathrm{lim}}}{V_{\mathrm{PT}}} \quad \text{and}$$

$$V_{\mathrm{Csat}} = w_{\mathrm{c}} E_{\mathrm{sat}}.$$

the model equations reduce to

$$J_{\mathrm{VCi}} = G_{\mathrm{VCi}}(V_{\mathrm{Ci}} - V_{\mathrm{Csat}}), \tag{3.5.2-90}$$

$$\frac{x_{\mathrm{t}}}{w_{\mathrm{c}}} = \frac{k}{k-1}\left(1 - \frac{J_{\mathrm{lim}}}{J_{\mathrm{C}}}\right) - \sqrt{\left(1 - \frac{J_{\mathrm{lim}}}{J_{\mathrm{C}}}\right)^2 - \frac{k}{k-1}\left(1 - \frac{J_{\mathrm{VCi}} + J_{\mathrm{lim}}}{J_{\mathrm{C}}}\right)} \quad \text{and} \tag{3.5.2-91}$$

$$\tau_{\mathrm{BCv}} = \tau_{\mathrm{c0}}\left[\left(1 - \frac{x_{\mathrm{t}}}{w_{\mathrm{c}}}\right)^2 + \frac{1}{k}\left\{2\frac{x_{\mathrm{t}}}{w_{\mathrm{c}}} - \left(\frac{x_{\mathrm{t}}}{w_{\mathrm{c}}}\right)^2\right\}\right]. \tag{3.5.2-92}$$

The model is valid for $J_{\mathrm{NDM}} \leq J_{\mathrm{C}} \leq J_{\mathrm{CK}}$, with the critical current density J_{CK}. Unfortunately, the system of equations cannot be solved for the critical current; the classical HICUM equations remain in place. The current density determining the onset of the NDM effect is voltage dependent via

$$J_{NDM}(V_{\mathrm{Ci}}) = q v_{\mathrm{sat}}\left[\frac{2\varepsilon}{q w_{\mathrm{c}}^2}(V_{\mathrm{Ci}} - w_{\mathrm{c}} E_{\mathrm{sat}}) + N_{\mathrm{C}}\right]. \tag{3.5.2-93}$$

An example for the application of the model is shown in fig. 3.20, where the unphysical low-current values were removed.

Remaining issues for an implementation into a compact model at this point are numerical conditioning of the equations (e.g. the pole in (3.5.2-91) for $J_{\mathrm{C}} = 0$) and the integration of (3.5.2-92) to obtain the collector charge. However, a more important issue occurs due to the assumed velocity step function: due to the reduced space charge in the region $x < x_{\mathrm{t}}$, the electric field is not inverted, as assumed for the calculation, violating the condition that $|E(x < x_{\mathrm{t}})| \leq |E_{\mathrm{sat}}|$, as shown in fig. 3.21. While the resulting equations are still correct assuming $v_{\mathrm{n}}(x < x_{\mathrm{t}}) = v_{\mathrm{sat}}$ (no assumption regarding $E(x)$ was made), this calls the physical validity of the model into question. For this reason, and because the combination of (3.5.2-91) and (3.5.2-92) is very complicated for implementation in a compact model, the step function is not considered further here.

Table 3.6: Parameters used for the application of the step-function model in fig. 3.20.

w_c	N_C	v_p	v_sat	E_p	E_sat
/μm	/(cm³)	/(cm s⁻¹)	/(cm s⁻¹)	/(kV cm⁻¹)	/(kV cm⁻¹)
0.2	$3 \cdot 10^{16}$	$3 \cdot 10^7$	$1 \cdot 10^7$	5	20

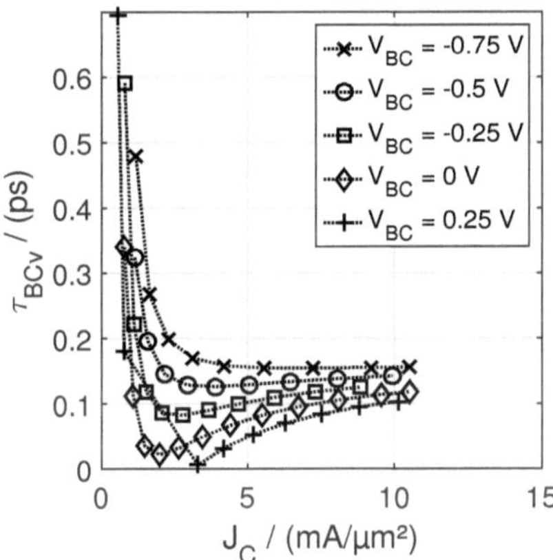

Figure 3.20: Results of (3.5.2-92) for InP collector structure with parameters as listed in tab. 3.6.

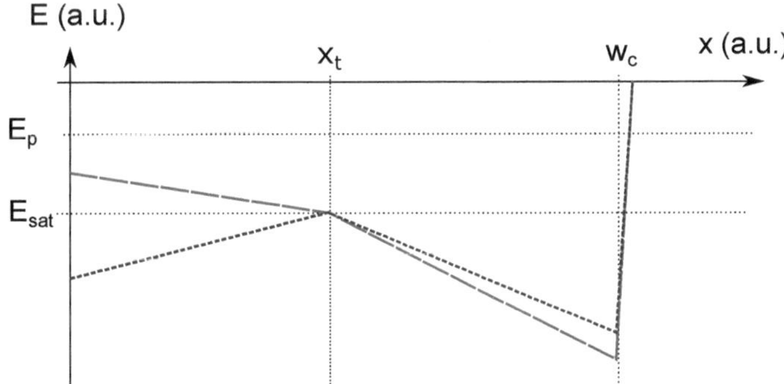

Figure 3.21: Electric field in the collector assumed for determining the velocity distribution (long dashed line) and actual electric field resulting from the velocity distribution (short dashed line) due to reduced space charge.

3.5.2.5 AHBT model approach

Existing III-V HBT models already take the NDM effect into account. Equations in the UCSD and AHBT models are based on the empirical model in [IRS$^+$03], with the AHBT adding terms for computational robustness [Agi16]. The complete model for the collector transit time reads

$$
\tau_c = \frac{1}{2}\left\{ \tau_{fc0}[1 - V_{\text{TC0Inv}} \cdot tf(V_{\text{BCi}}, V_{\text{TR0}}, V_{\text{MX0}})] \right.
$$
$$
+ 2\tau_{cmin}[1 - V_{\text{TCMINInv}} \cdot tf(V_{\text{BCi}}, V_{\text{TRMIN}}, V_{\text{MXMIN}})] -
$$
$$
\left. \frac{\tau_{fc0}[1 - V_{\text{TC0Inv}} \cdot tf(V_{\text{BCi}}, V_{\text{TR0}}, V_{\text{MX0}})] \cdot [I_{cfq} - I_{\text{TC}}(1 - V_{\text{BCi}}V_{\text{TCInv}})]}{\sqrt{[I_{\text{TC}}(1 - V_{\text{BCi}}V_{\text{TCInv}}) - I_{cfq}]^2 + [I_{\text{TC2}}(1 - V_{\text{BCi}}V_{\text{TC2Inv}})]^2}} \right\},
$$

$$(3.5.2\text{-}94)$$

using the auxiliary function

$$
tf(x, xtr, xmax) = \frac{1}{2}\left(\sqrt{(x + x_{\max})^2 + x_{tr}^2} + x - x_{\max}\right) \tag{3.5.2-95}
$$

with the model parameters τ_{fc0}, τ_{cmin}, V_{TC0Inv}, V_{TR0}, V_{MX0}, V_{TCMINInv}, V_{TRMIN}, V_{MXMIN}, I_{TC}, I_{TC2}, V_{TCInv} and V_{TC2Inv}. I_{cfq} is a modified version of the collector current for stability, though not explicitly described in the model equations. The model contains a (current-independent) high-current collector transit time τ_{cmin}, which is physically accurate, since $\Delta\tau_{\text{BCv}}$ typically does not become zero, but not considered in this work since it is impossible to separate it from the base transit time.

Setting τ_{cmin} to 0, determining τ_{fc0}, V_{TC0Inv}, V_{TR0} and V_{MX0} from the BC voltage dependence of the transit time at low current and V_{TCInv}, V_{TC2Inv}, I_{TC} and I_{TC2} from the current dependence yields the results shown in fig. 3.22. The model reproduces the general shape of the transit time well, but does not fit ideally in the transition region. A possible reason for this is the automated parameter extraction; however, given the large parameter number, the automation cannot be avoided.

Fig. 3.22a shows a problem of the model: The transit time is not limited to positive values with rising V_{BC}. This may result in negative transit times, especially during large-signal bias sweeps. While the problem is more visible here due to $\tau_{cmin} = 0$, it can occur for all parameter sets.

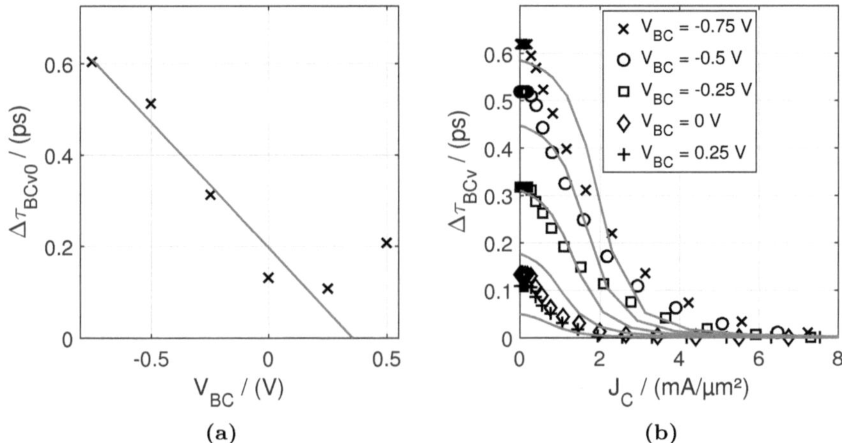

Figure 3.22: (a) V_{BC} dependence of the collector transit time at $J_C = 0$ and (b) current dependence of the collector transit time at various V_{BC}. Symbols: measurement, lines: model according to (3.5.2-94).

3.5.2.6 Discussion

Two physics-based approaches and two (semi-)empirical approaches towards modeling the current-dependent collector transit time have been presented.

The two physics-based approaches - the linear inverse velocity assumption and the velocity step function - yield results that are roughly similar to the measured characteristic. However, for the linear inverse velocity, the agreement with the measured data is not very good due to the simplifying assumptions that are necessary to obtain integrable transit time equations. Furthermore, the internal physical consistency of the step function model is questionable due to the violation of the assumptions regarding the electric field. Despite strong simplifying assumptions, the obtained equations are quite complicated and not very suitable for integration in a compact model. The two approaches demonstrate that, due to the relation of the electric field, the carrier velocity and the space charge, a fully physics-based approach for the current-dependent collector transit time is extremely difficult to find and implement.

On the other hand, the two (semi-)empirical models show good or very good agreement with measured data. Of the two, the AHBT model has the disadvantage of a very high model parameter number, the function of which is not immediately apparent from the equations; given that it is difficult to obtain measured data for τ_c, it seems more advantageous to use as low a parameter number as possible. Furthermore, the model is not properly limited for positive V_{BC}.

Therefore, the empirical model for the collector current dependence of the transit time as presented in section 3.5.2.1, using the physics-based voltage dependence as described in section 3.5.1.4 is the most suitable available approach for compact modeling.

CHAPTER **4**

Compact model parameter determination

4.1 Introduction

Even the best compact model is useless without an accurate parameter determination. For a physics-based compact model, this becomes even more important, since the relation of model parameters to certain physical effects means that those effects must be separated clearly. Otherwise, the advantages of physics-based modeling, such as scalability and predictive capability, cannot be exploited. Modern compact models can have upwards of 100 parameters, many of which describe physical effects that cannot easily be separated, but may have different frequency or temperature dependencies. Parameter determination is therefore non-trivial. When trying to determine model parameters by curve fitting, this complexity usually leads to a bad agreement between model and measurement beyond the range of the available fit data if a result is obtained at all.

Section 4.2.1 describes the parameter extraction flow in general, illustrating the most important extraction steps and their interdependence in order to supply a basis for the evaluation of parameter extraction methods and the model application. The first step towards parameter determination are measurements. Error-affected measurements necessarily lead to a bad compact model, since the error sources will be included in the transistor model. Therefore, sections 4.2.2 and 4.2.3 discuss typical DC- and RF- as well as pulsed on-wafer measurements.

Extraction methods for the series resistances have been published in the literature for many years. The resistances impact nearly all transistor characteristics in the regions important to circuit design. Their extraction is one of the first steps in the extraction flow, and a wrong value can impact subsequently extracted parameters. Extraction methods often make assumptions about an equivalent circuit or neglect certain effects in the devices. It is uncertain whether these simplifications are valid. Therefore, the extraction method evaluation for III-V HBTs is a major focus of this work. It is described in section 4.3.

Section 4.4 gives an example for a parameter extraction based on measurements of a single transistor, since this is the most likely realistic case for III-V HBTs at the time of this writing. The extraction steps are explained in some detail and differences between SiGe and III-V HBTs are highlighted where necessary.

Ideally, each model parameter can be extracted independently and uniquely. While this is not always possible, certain regions and layers of the devices can be fabricated independently or with additional measurement contacts. These test structures allow gathering information that would normally not be accessible. This supports the creation of geometry-scalable models, which is demonstrated in section 4.5, with a focus on test structures as they can be manufactured in III-V processes in section 4.5.2 and the separation of transistor properties into internal and external elements in section 4.5.3.

4.2 Parameter determination

4.2.1 Parameter determination methodology

In a physics-based compact model like HICUM, the extraction of parameters is complicated by the interdependence of various physical effects unless the proper extraction flow is applied. An example of the typical SiGe extraction flow is given in fig. 4.1 [Kra15]. It is also applicable to III-V HBTs.

In the beginning of the extraction, a reference device is chosen; all non-scaling parameters are determined for this device. The junctions capacitance parameters and the parasitic capacitances are determined from suitable test structures or from small-signal device measurements. They are separated into internal and external resp. peripheral parts based on multiple measured device geometries, if available. The series resistances are determined either from extraction methods or from test structures or estimated from structure and material information (sheet resistances) or literature information (contact resistances). The junction diode parameters can be determined from low-bias device measurements and separated in a similar way.

Given measurements at different ambient temperatures, the above steps can be repeated for each temperature. Since all extracted elements so far can be determined at low bias, self-heating is not relevant yet. In this way, the temperature dependence of the series resistances, junction diodes and capacitances can be extracted. Having obtained those, it is possible to determine the thermal resistance, which is necessary at this point because of the impact of self-heating on subsequent extraction steps.

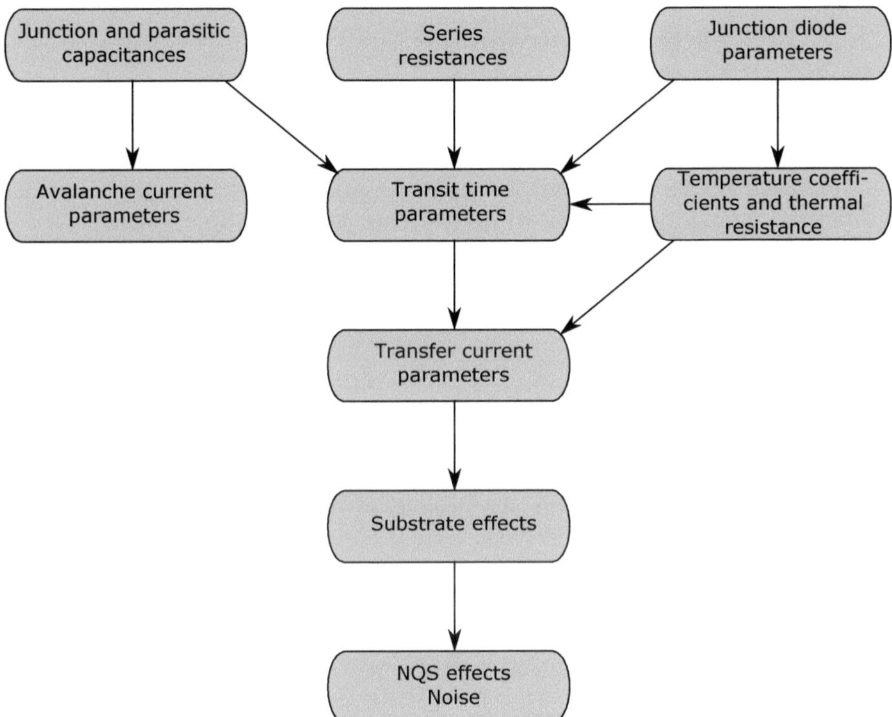

Figure 4.1: Extraction flow for HICUM/L2.

The avalanche current parameters are extracted from device measurements in regions in which self-heating is irrelevant. They require a previous extraction of the junction capacitance parameters since the electric field in the BC SCR is modeled via the normalized BC capacitance.

In charge-based models, the transfer current depends, among others, on the diffusion charge. Therefore, the transit time parameters need to be extracted before the transfer current parameters. This is possible because the transit time depends on the transfer current, but not on the transfer current parameters, so the measured current can be used directly during this step.

With all charge-related parameters known, the transfer current parameters can be extracted based on DC measurements of the reference device.

Finally, the parasitic substrate transistor and substrate network are modeled based on reference device measurements, and the non-quasi-static parameters are determined based on high-frequency measurements, if available. Similarly, if noise measurements have been taken, the flicker noise model parameters can be extracted.

4.2.2   On-wafer measurements

Device characterization for compact modeling is the most critical type of measurement where accuracy is concerned, since all errors here are translated into all circuits based on the model generated from those measurements.

Measurements for compacts models are commonly taken directly on-wafer, without packaging and usually without dicing the wafer first. The reason is that compact models are usually intended for integrated circuits, not discrete components, since they are used when high performance is required. Additionally, for both scalable and statistical models, many devices need to be measured, which is easier with an automated on-wafer setup.

On-wafer measurements offer the advantage that no package parasitics need to be taken into account, but a well-defined method of contacting the device-under-test (DUT) is needed nonetheless. Therefore, transistors are usually fabricated in ground-signal-ground (GSG) pads, where the signal-carrying probe tip is surrounded by ground pads and electromagnetic interference is suppressed. Typically, the signal-carrying pads are the base and the collector, with the emitter and substrate being grounded. Additionally, small-signal measurements must be both calibrated and deembedded, while for DC measurements, the user needs to make sure that no oscillations occur in the measurement setup and that contact resistances from the probes to the wafer are removed.

It is preferable to use a standard setup with well known equipment for device characterization, but for many laboratories, this is not practical due to the different required measurement setups for a full device characterization:

- Standard setup for two-port small signal measurements, typically in the range from 500 MHz to 50 GHz. Different probes may be required for different technologies on a temperature-controlled wafer prober. This is the only setup required for single-transistor extraction.

- High-frequency setups for measurements up to 110, 220 or even 325 GHz, each requiring different hardware.

- Standard DC setup with up to six DC sources and probes.

- Large-signal setup.

- Noise measurement setup.

Note, that not all of those measurements are required for parameter extraction. High-frequency, large-signal and noise measurements primarily serve as verification for the model, though each can supply useful additional information if available, and a DC setup is only required if suitable test structures are available (cf. section 4.5.2).

Calibration is the process of determining the systematic error of a measurement and removing it from the measured data. Vector network analyzers measure the small-signal parameters of a DUT. Since they measure not just the magnitude, but also the phase of the signal with a high dynamic range and the high demands on the accuracy of both, they require regular calibration. Standard calibration routines are often supplied in the VNA software and applied directly to the measured data, but output of the raw data is also possible if the user prefers it. Regular, usually daily, recalibration is required due to drift of errors even if no components in the measurement setup have changed. Many different calibration methods are available in the literature. They differ in the use of calibration standards, the effort of implementation and in their accuracy over frequency [WCS+14, RSD+11]. In this work, short-open-load-thru (SOLT) and load-reflect-reflect-match (LRRM) calibration were used and little difference in the frequency range up to 50 GHz was found between both.

Deembedding is the process of removing the impact of the contact pads and lines that are not part of the DUT and would not be included when using it in a circuit. This is accomplished by measuring so-called dummy structures, one with an open connection and one with a short connection where the DUT would ordinarily be located, and removing the unwanted elements determined in this way from the actual DUT measurement. For RF measurements or when very precise data are needed, more complicated deembedding may be necessary.

Tab. 4.1 lists the typical required measurements for parameter extraction, though details depend on the used extraction methods and technology behavior. All measurements need to be taken at at least three different ambient temperatures in order to determine the temperature dependence parameters. If a geometry-based separation is intended, all measurements also need to be performed for at least three different devices. A practical application of the extraction flow along with notes specifically for III-V based HBTs can be found in section 4.4.

Table 4.1: Required measurements of a single device in GSG pads for parameter extraction.

No	Measurement	Data	Extr. parameters
1	DC forward transfer characteristic at multiple V_{BC}	$V_{\mathrm{BE}}, V_{\mathrm{BC}}, I_{\mathrm{C}}, I_{\mathrm{B}}$	BE junction diode transfer current
2	DC reverse transfer characteristic at $V_{\mathrm{BE}} = 0$ V	$V_{\mathrm{BC}}, I_{\mathrm{C}}, I_{\mathrm{B}}$	BC junction diode
3	DC sweep to negative V_{BC} at $V_{\mathrm{BE}} = 0$V	$V_{\mathrm{BC}}, I_{\mathrm{C}}, I_{\mathrm{B}}$	BC leakage/tunneling
3	DC sweep to negative V_{BE} at $V_{\mathrm{BC}} = 0$V	$V_{\mathrm{BE}}, I_{\mathrm{C}}, I_{\mathrm{B}}$	substrate diode substrate transistor
4	DC forward output characteristic at multiple constant V_{BE}, including several very low values for V_{BE}	$V_{\mathrm{BE}}, V_{\mathrm{BC}}, I_{\mathrm{C}}, I_{\mathrm{B}}$	avalanche current R_{th} verification
5	DC forward output characteristic at multiple constant I_{B}	$V_{\mathrm{BE}}, V_{\mathrm{BC}}, I_{\mathrm{C}}, I_{\mathrm{B}}$	R_{th} verification
6	Cold S-parameters, V_{BE} and frequency sweep, at $V_{\mathrm{BC}} = 0$ V	$V_{\mathrm{BE}}, \underline{S}$	BE junction and parasitic capacitances
7	Cold S-parameters, V_{BC} and frequency sweep, at $V_{\mathrm{BE}} = 0$ V	$V_{\mathrm{BC}}, \underline{S}$	BC and CS junction and BC parasitic capacitances
8	Hot S-parameters, V_{BE} and frequency sweep, at multiple V_{BC}	$V_{\mathrm{BE}}, V_{\mathrm{BC}}, I_{\mathrm{C}}, I_{\mathrm{B}}, \underline{S}$	Transit time high-frequency

4.2.3 Pulsed measurements

Pulsed measurements are DC or small signal measurements that are taken in a very short time interval during which the device is on. The device is turned off nearly immediately before and after the measurement. The purpose of pulsed measurements for HBTs is to eliminate or reduce the impact of self-heating. For other devices, e.g. CNTFETs, pulsed measurements can be used to investigate charge trapping [HCN$^+$14].

Since the measurement range for III-V HBTs, especially InP HBTs, is limited by thermal breakdown of the devices, and since self-heating usually makes parameter extraction more difficult, pulsed measurements were investigated as a possible tool for modeling in this work. This investigation is based on a Auriga AU 4550 pulsed IV system coupled with an Agilent PNA. The Auriga system has a minimum pulse duration of 300 ns [Aur09].

Fig. 4.2 shows the schematic shape of a voltage pulse used for measurement. Important parameters are the pulse width (t_{w}) and the pulse repetition interval (t_{ri}). The duty cycle P_{DC} of the measurement can be calculated as

$$P_{\mathrm{DC}} = \frac{t_{\mathrm{w}}}{t_{\mathrm{ri}}}, \tag{4.2.3-1}$$

and the pulse repetition frequency f_{ri} is the inverse of the pulse repetition interval.

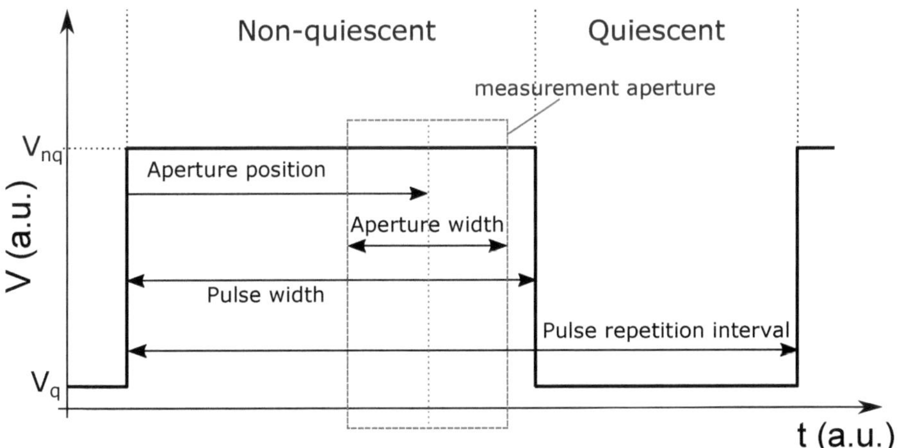

Figure 4.2: Schematic representation of pulsed measurement input or output voltage over time. Several definitions ar marked in the figure.

The measurement aperture, i.e. the timeframe in which the measurement data is acquired, is indicated in the figure by a dashed rectangle. The user can change the aperture position within the pulse and the aperture width. Both must be chosen so that minimal self-heating takes place while repeatable data is obtained.

Pulsed RF measurements can be taken in wideband and narrowband mode, depending on the capabilities of the VNA and the settings of the pulsed IV system. In order to understand the difference, visualizing the resulting spectrum of the superposition of the rectangular pulse signal and the sinusoidal AC signal is helpful.

Fig. 4.3 shows the time-domain and the resulting frequency-domain signal. Despite the narrowband mode occasionally being referred to as continuous wave (CW) measurement since the network analyzer signal is continuous, but in both cases, the receiver evaluates the AC signal only during the assigned measurement window in the DC pulse non-quiescent state. This results in a spectrum with discrete components with a distance given by f_{ri} and a power given by an envelope $sinc(f)$ function, the width of which is determined by t_{w} [Agi07], centered around the measurement frequency f_0. In order to measure the S-Parameters, i.e. the ratio of incident to reflected resp. transmitted power, correctly, the receiver must either take the entire spectrum into account (wideband mode) or filter one frequency component without interference from others (narrowband mode).

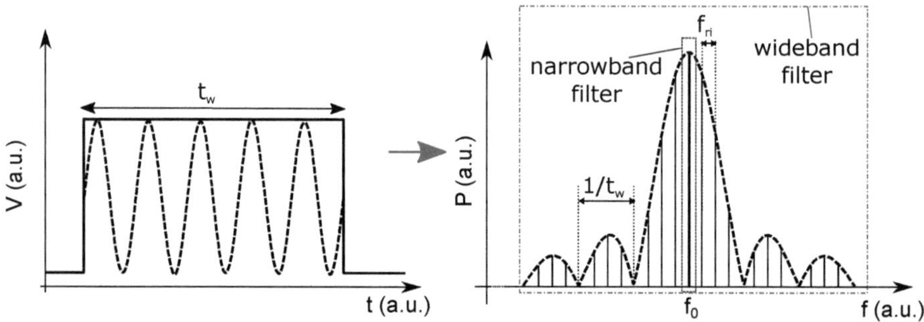

Figure 4.3: Schematic representation of pulsed small signal parameter measurement in time and frequency domain.

Therefore, measurements in narrowband mode must have a sufficiently large f_{ri} resp. a sufficiently short t_{ri} or high duty cycle to filter the center frequency only. The disadvantage of this approach is that part of the signal is lost and therefore the dynamic range of the VNA is reduced. On the other hand, in wideband mode, arbitrary pule repetition frequencies are permissible, but a too short pulse width may spread the spectrum beyond the bandwidth of the receiver. For the short pulses desired for quasi-isothermal HBT measurements, the narrowband mode is more suitable.

In order to investigate the suitability of the pulsed setup for III-V measurements, multiple values for pulse width and measurement aperture were investigated. Furthermore, the actual pulse shape was measured using an external oscilloscope after some results were impossible to explain based on internal data from the pulsed IV system alone.

Fig. 4.4 shows the impact of the aperture position and the pulse width on the measured data. According to fig. 4.4a, the aperture position has little impact on the acquired data. Note, that the aperture width for this figure is $P_{\mathrm{AW}} = 500\,\mathrm{ns}$, except for $P_{\mathrm{AP}} = 200\,\mathrm{ns}$, where it is $P_{\mathrm{AW}} = 125\,\mathrm{ns}$ in order to avoid measuring part of the quiescent region. Notably, there appears to be no difference between the four measurements in the later region of the pulse; it is concluded that the transistor has reached maximum self-heating already. Further, an early measurement and short aperture leads to measurement errors, since a much lower current is measured here even in regions in which self-heating is not relevant. The minimum aperture width for correct DC measurements is therefore $P_{\mathrm{AW}} > 125\,\mathrm{ns}$.

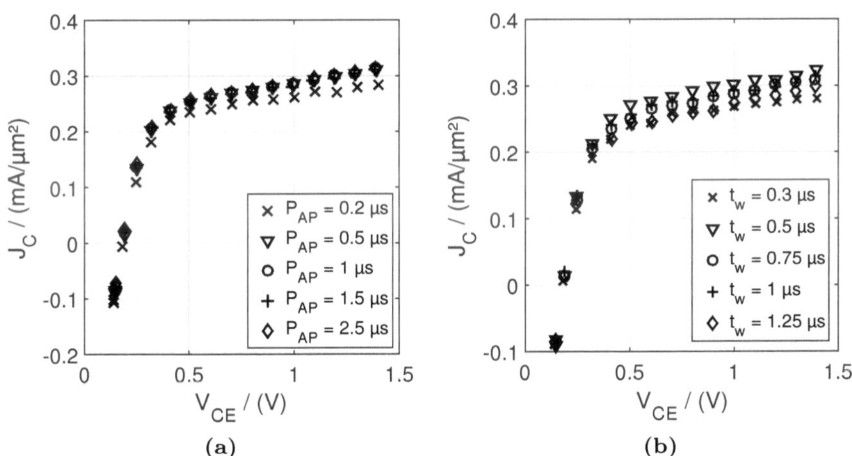

Figure 4.4: Forward output characteristic for an InP HBT during pulsed measurement with constant V_{BE}; (a) changing aperture position within pulse and (b) changing pulse width with constant aperture position relative to end of pulse.

Fig. 4.4b shows the measured current for different pulse widths. The aperture for this measurement is approximately $P_{\mathrm{AP}} = 0.5t_{\mathrm{w}}$ and $P_{\mathrm{AW}} = (0.4\ldots0.6)t_{\mathrm{w}}$. Unexpectedly, with the exception of the shortest pulse width with a very short aperture, the current density decreases with pulse and aperture length. There is no physical reason for this behavior immediately apparent. Therefore, the pulse shape is investigated using an external oscilloscope.

Fig. 4.5 shows the pulse shape with both a reference 50 Ω resistance and with the DUT connected. In both cases, it is apparent that no measurement can be taken for approximately 300 ns after the beginning of the pulse because of the unstable input signal. Combined with the previously mentioned minimum required aperture width, this severely limits the possibility of isothermal measurements; while self-heating may be partially preventable, is cannot be ignored entirely.

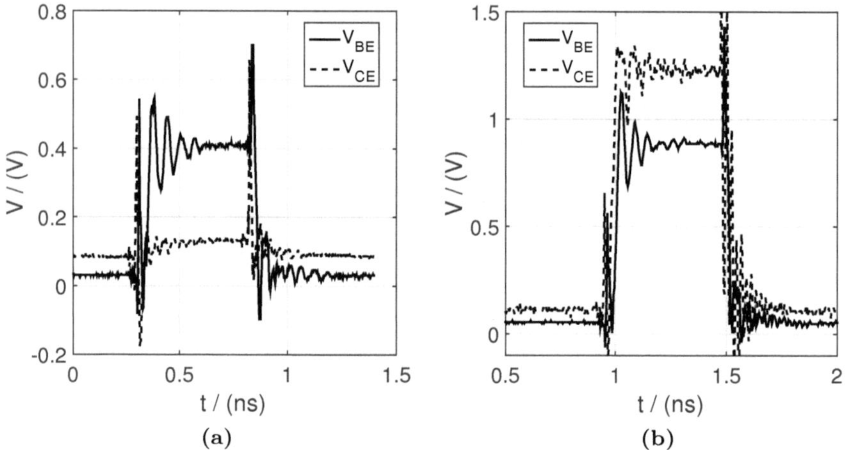

(a)   (b)

Figure 4.5: Pulsed IV system pulse shape over time, measured by external oscilloscope, (a) with 50 Ω calibration resistance attached and (b) connected to transistor.

A second important point is the pulse stability over time. For 4.6 shows the externally measured pulse for a pulse width of $t_\mathrm{w} = 1\,\mu\mathrm{s}$. While at first glance the voltage in fig. 4.6a appears stable, in the magnified view in fig. 4.6b, a variation of approximately $\Delta V = 10\,\mathrm{mV}$ is visible. Since the transfer current in a HBT depends exponentially on the input voltage, this leads to a current variation of approximately 50 %. The current is, however, not directly measurable with the oscilloscope and cannot be shown here.

Pulsed S-Parameter measurements were attempted as well. The results for multiple pulse conditions are shown in fig. 4.7. The measured transit frequency agrees well with the value from standard RF measurements, though the pulsed data is very noisy. The deviation at low bias can be explained with

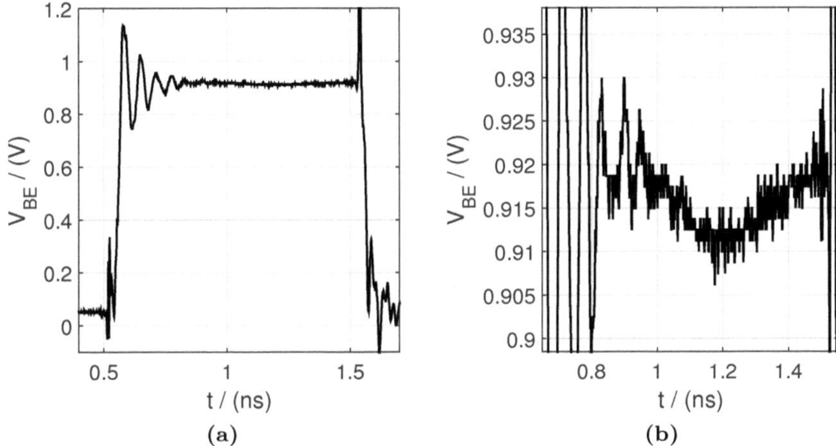

Figure 4.6: Pulsed IV system pulse shape over time, measured by external oscilloscope, (a) connected to DUT for long pulse duration and (b) magnification of relevant region of (a).

the fact that the regular S-Parameter measurement was taken on the HP8510, which is less sensitive than the PNA. A significant deviation in the medium to high bias range is visible only for short apertures. In order to investigate whether this deviation is a measurement artifact, $\mathrm{Im}(\underline{Y}_{11})$ is also shown. It is clearly visible that the pulsed measurement indicates a large increase in the Y-Parameter, which is not physically meaningful.

Summarizing the investigation into pulsed DC and S-Parameter measurements, this method is not employed in this work because

- the long time it takes for the pulse to settle in combination with the relatively long required P_{AW} for consistent measurement results means that the benefit of avoiding self-heating is negligible.

- small signal data taken during pulsed RF measurements are noisy.

- pulse stability over time is not guaranteed.

Given improved equipment, those points would have to be revisited, but with the pulse system available at the time of this writing, isothermal or quasi-isothermal DC and RF pulsed measurements are impossible. This also illustrates the necessity to check measurement results and equipment before modeling.

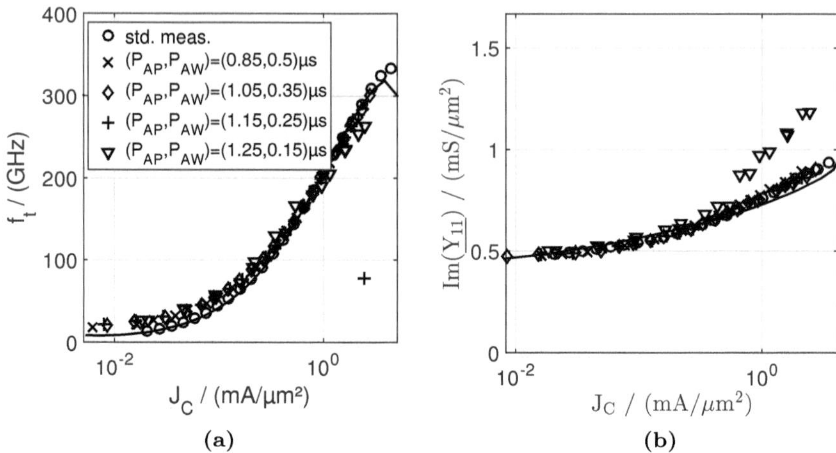

Figure 4.7: (a) f_t from regular and various pulsed measurements (symbols) as well as compact model (solid line). (b) Comparison of $\mathrm{Im}(\underline{Y}_{11})$ from regular and pulsed measurement. The legend in (a) is valid for both plots.

4.3 Evaluation of extraction methods for series resistances

4.3.1 Evaluating extraction methods

It was already established that, in order to design and optimize circuits that exploit the peak performance of a given HBT technology, physics-based geometry scalable compact models are required due to their predictive capability. For instance, their parameters can be determined at relatively low frequencies while retaining accuracy up to frequencies where experimental characterization is difficult for small-signal and impossible for large-signal operation [SP14]. A major obstacle for the widespread deployment of such compact model is the difficulty of the parameter extraction. A unique and reliable determination of parameters simplifies the extraction effort greatly and increases the reliability of compact models [Paw14, Kra15]. The evaluation of extraction methods presented in this sections aims, in part, at improving the overall compact model quality for III-V HBTs.

As fig. 4.1 shows, the extraction of the series resistances is one of the first steps in the extraction flow. This is because they affect many of the subsequent steps, and a wrong value for any of them causes errors in other parameters or leads to the model being unable to reproduce the measured characteristics.

Series resistances are usually determined in one of three ways:

- From test structures and known device geometry, as described in section 4.5.2.

- Calculated from knowledge of the device structure and material parameters.

- Determined from terminal-data based extraction methods, i.e. from DC- or small-signal data.

The latter are often the only option during single-device extraction, especially when the modeling engineer has little or no contact with the foundry. Furthermore, it is generally very difficult or impossible to create test structures for the emitter resistance, and even if the device structure is known very well, the contact resistances are still unknown; therefore, terminal-based extraction methods often remain the only option.

Series resistance extraction methods are applied to measurements, using simplified relationships between voltages, currents and S-parameters to determine the target resistance value; though they sometimes require previous extraction steps to be performed, they rarely or never require additional information about the device structure. Typically, the device is operated in a region (or the measured data are extrapolated into a hypothetical operating region) in which a simplified equivalent circuit is applicable compared to those provided in PDKs and used for circuit design and a single series resistance dominates the measured characteristic either directly or can be determined from a relatively simple calculation. For this purpose, some assumptions about negligible elements in the EC or the device behavior are made. These assumptions may be valid only for certain technologies and less reliable for others. Fig. 4.8 shows the equivalent circuits typically used for the extraction methods investigated here.

Verifying an extraction method leads to the problem of the reference value determination. Often, newly published methods are used on a single transistor or several transistors of a single technology. The reference value against which the method quality is compared is taken from test structures or other extraction methods; alternatively, the verification is simply a good "looking" performance of a compact model into which the extracted value is inserted or by correctly extracting an added resistance. This is clearly insufficient. Beyond the problematic reference value determination, the correct reference value may even change with bias or be differently defined in a compact model due to its distributed nature (e.g. for a non-constant current flow in the

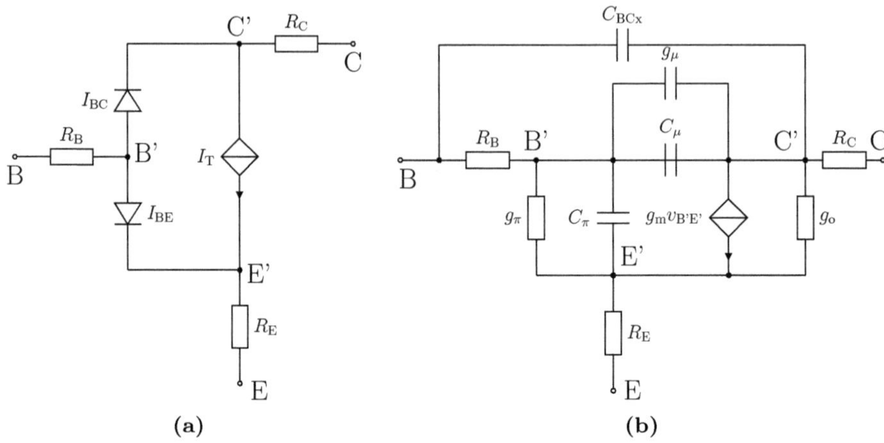

Figure 4.8: (a) DC and (b) small-signal equivalent circuit assumed for determining the series resistances by most methods considered in this work. For some methods, additional simplifications were made. The names listed in this figure are used in the analysis below and have their usual meanings.

direction of the resistance calculation due to a perpendicular flow across a junction, a lumped resistance can model either the voltage drop or the power dissipation correctly, but not both).

Therefore, in order to be able to evaluate the absolute (achievable) accuracy, in this study the extraction methods under investigation will be applied to simulation data generated by the physics-based compact model HICUM, which has been shown to accurately model InP DHBTs [NSC+13]. The advantages of this approach are:

- The bias-dependent target value is known so that the error of a method can be easily assessed, and

- the compact model allows the analysis of the limitations of a method by turning physical effects on or off and providing access to internal nodes and element values.

As a necessary precondition, the model needs to include all relevant physical effects and to accurately reproduce the measured data in the region of interest. Given that extraction methods assume lumped element equivalent circuits, applying them to compact model data represents a best-case scenario with regard to the distributed resistance as described above. Additionally, emitter current crowding, emitter perimeter injection and the split of R_B into its external and internal component and resistance changes due to self-heating are taken into account in HICUM. Note, that most compact models and several extraction methods differentiate between the bias-independent external part of the base resistance R_Bx and the internal, bias-dependent part R_Bi so that

$$R_\mathrm{B} = R_\mathrm{Bx} + R_\mathrm{Bi} \qquad (4.3.1\text{-}2)$$

while still using a simplified equivalent circuit with the lumped total base resistance. Typically, the two contributions are separated by extracting R_B for different bias points.

The list of extraction methods gathered in this work corresponds to a careful selection of the most promising methods, which have been applied in this study to the devices from different geometries and technologies described in section 4.3.2.

In this investigation, methods known to possess a flawed physical background such as the extraction of the base resistance from the collector current characteristic [Log72], have been omitted. The semi-insulating substrate of III-V technologies prevents the use of methods based on the substrate current flow [ZBB+96]. Methods based on impact ionization (e.g. [VCT+93]) are generally unsuitable for III-V devices since the high breakdown voltages caused by the high-bandgap collector material in combination with high current densities leads to a high probability of device destruction when attempting to take the required measurements [Zam16]. Elements corresponding to the substrate or avalanche current have therefore already been omitted from the equivalent circuits in fig. 4.8.

The results presented here were also published in part in [NKS14, NS15, NKPS16]. At the time of this writing, compact models were available only for InP HBTs; the discussion herein is therefore limited to those devices, but is expected to be at least partially applicable to, e.g., GaAs HBTs.

4.3.2 Investigated process technologies

A variety of InP HBT process technologies have been investigated. For each process, several devices were measured which differed in emitter mesa width (b_E0), length (l_E0) and area ($A_\mathrm{E0} = b_\mathrm{E0} \cdot l_\mathrm{E0}$). The investigated transistors are from a GCS process [GCS] with $A_\mathrm{E0} = \mathbf{0.8x(3, 5, 10)}$ $\mu\mathrm{m}^2$ as well as $A_\mathrm{E0} = \mathbf{(0.5, 0.8, 1, 1.5, 2)x15}$ $\mu\mathrm{m}^2$ from two different process runs, both with $(f_\mathrm{T}, f_\mathrm{max}) = (300, 250)$ GHz, a Teledyne process [GDR+05] with $A_\mathrm{E0} = 0.5x(3, 5, \mathbf{10})$ $\mu\mathrm{m}^2$ and $(f_\mathrm{T}, f_\mathrm{max}) = (300, 450)$ GHz and an FhG IAF process [DRM+11] with $A_\mathrm{E0} = 0.7x(2, 4, 12)$ $\mu\mathrm{m}^2$ and $(f_\mathrm{T}, f_\mathrm{max}) =$

$(300, 260)$ GHz. Dimensions marked in bold indicate transistor sizes for which a complete compact model was available. For the GCS transistors, both geometry scalable HICUM/L2 parameters were extracted using suitable test structures on a special test chip [NSC$^+$13] and a HICUM/L0 model was created based on single-transistor extraction methods [NSSL13]. For the two other technologies, only HICUM/L0 model parameters could be determined since only single-transistor structures were available. The HICUM model has been in use in the industry for many years and is capable of accurately modeling all available device sizes.

All data for the investigation presented here were generated and measured, respectively, at a chuck temperature of $T_0 = 300$ K. All S-parameter data are open-short deembedded; parasitic capacitances as mentioned in this work refer to the influence of vias and device contacts which cannot be deembedded in this fashion.

4.3.3 Methods for the emitter resistance

The emitter series resistance becomes relevant in the medium to high-current region or at high frequencies and is often difficult to separate from other effects, most notably the reverse Early effect and the self-heating of the device, but is very important for the DC operation of a transistor due to the negative feedback it causes for the input voltage in common-emitter operation.

Several methods exist that attempt to determine the emitter resistance from DC [HVvN04, RAH11, NT84, GR98, TSW$^+$97, PLS14] or low frequency [MNT92, GTB97, DB03] as well as from the high-frequency [SM72, NH91] terminal quantities. The most promising methods are investigated in this work.

Sections 4.3.3.1 - 4.3.3.7 discuss the methods in order to point out possible problems when applying them to III-V HBTs; since all methods have been published elsewhere, however, not all details are necessarily presented here. Section 4.3.3.8 presents the results of the method application to both compact model generated and measured data, summarizes the obtained results and gives application recommendations.

4.3.3.1 g_{mi} method

The method was published in detail in, e.g., [HVvN04], though this does not indicate the method origin, since older applications can be found (e.g. [BCS$^+$02]). It has been widely employed for the determination of the emitter resistance and is based on the assumption that any deviation of the measured transconductance g_{m} from an ideal extrapolated low-bias value is due to the series resistances.

From a simplified formulation of the collector current, taking the series resistances into account and writing the base current as $I_\mathrm{B} = I_\mathrm{C}/\beta_\mathrm{f}$ with the small-signal low-frequency current amplification β_f,

$$I_\mathrm{C} = I_\mathrm{S} \exp\left[\frac{V_\mathrm{BE} - I_\mathrm{C}(R_\mathrm{E} + (R_\mathrm{E} + R_\mathrm{B})/\beta_\mathrm{f})}{m_\mathrm{Cf} V_\mathrm{T}}\right], \qquad (4.3.3\text{-}3)$$

the external transconductance can be determined as

$$g_\mathrm{m} = \left.\frac{\partial I_\mathrm{C}}{\partial V_\mathrm{BE}}\right|_{V_\mathrm{CE}} = \frac{I_\mathrm{C}}{m_\mathrm{Cf} V_\mathrm{T}}\left[1 - g_\mathrm{m}\frac{(R_\mathrm{E} + (R_\mathrm{E} + R_\mathrm{B})}{\beta_\mathrm{f}})\right] \qquad (4.3.3\text{-}4)$$

when neglecting the current dependence of β_f, R_E and R_B.

Defining the internal transconductance

$$g_\mathrm{m} = \frac{\partial I_\mathrm{C}}{\partial V_\mathrm{B'E'}} = \frac{I_\mathrm{C}}{m_\mathrm{C} V_\mathrm{T}} \qquad (4.3.3\text{-}5)$$

allows to rewrite (4.3.3-4) as

$$\frac{1}{g_\mathrm{m}} = \frac{1}{g_\mathrm{mi}} + R_\mathrm{E}\left(1 + \frac{1}{\beta_\mathrm{f}}\right) + \frac{R_\mathrm{B}}{\beta_\mathrm{f}}. \qquad (4.3.3\text{-}6)$$

It is typically assumed that $m_\mathrm{Cf} = const.$ and $1/\beta_\mathrm{f} \ll 1$. Then, the emitter resistance can be obtained by linearly extrapolating $1/g_\mathrm{m} \approx 1/g_\mathrm{mi}$ from low current densities, where the voltage drop over the resistances is negligible, w.r.t. $1/I_\mathrm{C}$ from the intersection of the interpolation with the y-axis.

More advanced variants of the g_mi method consider a possible current-dependence of the nonideality factor as

$$m_\mathrm{Cf} = m_\mathrm{Cf0} + \frac{I_\mathrm{C}}{I_\mathrm{m}}, \qquad (4.3.3\text{-}7)$$

where m_Cf0 is the low-bias nonideality factor and I_m is a fitting term for the current-dependent increase, resulting in the correction resistance

$$R_\mathrm{cor} = \frac{V_\mathrm{T}}{I_\mathrm{m}}. \qquad (4.3.3\text{-}8)$$

Furthermore, the finite current amplification is taken into account by calculating the emitter resistance from the y-axis intercept R as

$$R_\mathrm{E} = \frac{R - (R_\mathrm{cor} + R_\mathrm{B}/\beta_\mathrm{f})}{1 + 1/\beta_\mathrm{f}}. \qquad (4.3.3\text{-}9)$$

In [HCZA96], a correction for the self-heating of the device was published by taking the temperature change caused by the current increase into account during the calculating of the transconductance as

$$\frac{1}{g_{\mathrm{m}}} = \frac{\mathrm{d}V_{\mathrm{BE}}}{\mathrm{d}I_{\mathrm{C}}} = \left.\frac{\partial V_{\mathrm{BE}}}{\partial I_{\mathrm{C}}}\right|_{T} + \left.\frac{\partial V_{\mathrm{BE}}}{\partial T}\right|_{I_{C}} \frac{\mathrm{d}T}{\mathrm{d}I_{\mathrm{C}}}. \tag{4.3.3-10}$$

This results in an additional correction resistance

$$R_{\mathrm{cor,T}} = \alpha_{\mathrm{VBE}} R_{\mathrm{th}} V_{\mathrm{CE}}, \tag{4.3.3-11}$$

with the thermal resistance R_{th} and the slope of the BE voltage w.r.t. the temperature at a constant collector current α_{VBE}.

An example of the application of this method both with and without the correction for high self-heating is shown in fig. 4.9. Note, that the transconductance can be determined from the DC derivatives or from $\mathrm{Re}(\underline{Y}_{21})$. The latter is preferred and was chosen here as it circumvents the impact of self-heating at least partially due to the high thermal time constant of the HBTs compared to typical S-parameter measurement frequencies (c.f. section 4.2.3). Then, only the DC operating point has an impact via V_{T}.

Overall, the g_{mi} method is not very well suited for InP DHBTs. Due to their comparatively low current amplification, the correction for the base resistance must be taken into account. Even when including it, the method performs relatively poorly; the most important error sources include

- self-heating, which is not sufficiently corrected for with the equations in [HCZA96];

- the dependence of β_{f} on the bias conditions, which was not taken into account in the method derivation and

- a required precisely known value for R_{B} and potentially its bias dependence due to the low current amplification.

4.3.3.2 Z-parameter method

The Z-parameter method was originally proposed in [MT92] and further described in [GTB97]. It is based on the assumption that the junction and parasitic capacitances can be neglected at sufficiently low frequencies and high current densities. Then, the real part of $\underline{Z}_{12}$ can be approximated as

$$\mathrm{Re}(\underline{Z}_{12}) = \frac{1}{g_{\mathrm{m}}} + R_{\mathrm{E}}. \tag{4.3.3-12}$$

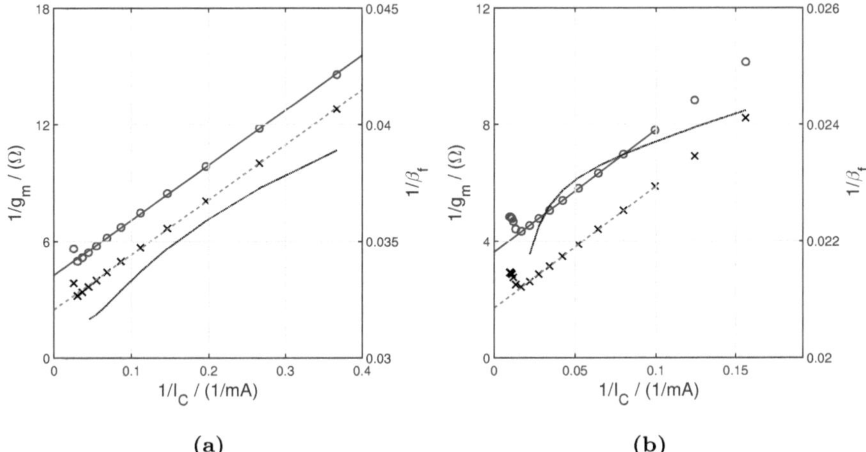

(a)   (b)

Figure 4.9: Application of the g_{mi} method to compact model data (symbols) via interpolation (lines) based on (a) a GCS device with $A_{\mathrm{E0}} = 15x0.8\,\mathrm{\mu m}^2$ and (b) a TD device with device with $A_{\mathrm{E0}} = 10x0.5\,\mathrm{\mu m}^2$. Solid line, circles: Including self-heating correction according to [HCZA96]. Dashed line, crosses: Without correction. The dotted line shows the inverse current amplification on the r.h.s. y-axis, indicating the impact of the base resistance on the extraction result.

Since $1/g_{\mathrm{m}} \propto 1/I_{\mathrm{C}}$, the emitter resistance can be determined from the extrapolating of $Re(\underline{Z}_{12})$ w.r.t. $1/I_{\mathrm{C}}$ towards infinite currents from the medium to high current density region. The emitter resistance is given by the intercept R of the extrapolation with the y-axis. Note, that the extraction was applied in [GTB97] w.r.t. $1/I_{\mathrm{B}}$, which is mathematically incorrect.

A potential error of the method is that, even including the assumptions of infinite conductances g_{pi} and g_{m} in the limit of infinite I_{C}, a thorough analysis of the equivalent circuit reveals that (4.3.3-12) is incorrect and $Re(\underline{Z}_{21})$ contains other contributions. A correction was published in [DB03] as

$$R \approx R_{\mathrm{E}} + R_{\Delta} = R_{\mathrm{E}} + \frac{(1 - \alpha_0)C_{\mathrm{BCx}}R_{\mathrm{B}}}{C_{\mathrm{BCx}} + C_{\mu}} \qquad (4.3.3\text{-}13)$$

and applied here. Note, that the correction is based on the T equivalent circuit, not the π equivalent circuit used by HICUM, and α_0 is defined as

$$\alpha_0 = \frac{I_{\mathrm{C}}}{I_{\mathrm{E}}} \approx (1 - \frac{1}{\beta_{\mathrm{f}}}). \qquad (4.3.3\text{-}14)$$

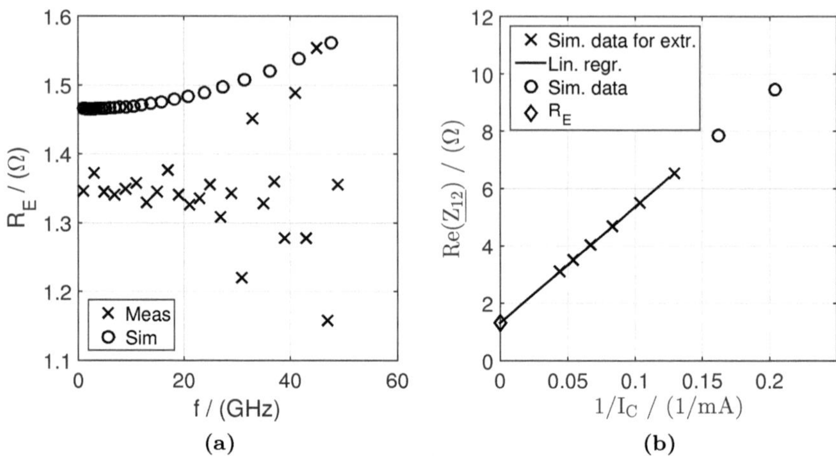

Figure 4.10: (a) Frequency dependence of extracted value for the Z-parameter method for compact model generated and measured data and (b) application of the Z-parameter method to compact model data (symbols) via interpolation (line) based on a GCS device with $A_{\mathrm{E0}} = 15x0.8$ µm^2. The intercept of the interpolation with the y-axis indicates the extracted value for R_E (diamond). Crosses show the chosen extraction range, circles show other data.

The extraction frequency range for both simulated and measured data was investigated in fig. 4.10a. The Z-parameter method is only weakly frequency-dependent. An example for the extraction is shown in fig. 4.10b.

Errors for this method may be caused by

- the dependence of the extracted value on the extraction region, since the linear relation of $Re(\underline{Z_{12}})$ vs. $1/I_\mathrm{C}$ can be difficult to define;

- the overly simplifying assumptions about the equivalent circuit elements, e.g. neglecting the current flow through the capacitances entirely;

- the required knowledge about the capacitance split across R_B the and the exact value of the base resistance if applying the correction (4.3.3-13);

- the self-heating of the device causing a nonlinear relationship between $1/I_\mathrm{C}$ and $1/g_\mathrm{m}$.

In order to investigate the method in detail, a full analytical solution for the Y- and Z-Parameters of the HICUM/L2 equivalent circuit was obtained.

Then, all terms with a negligible influence on the final result for $\mathrm{Re}(\underline{Z}_{11})$ were removed based on a compact model simulation and evaluation of individual terms. Finally, an extrapolation towards infinite current performed by setting $g_{\mathrm{m}} = \infty$. Then, the remaining terms are

$$\mathrm{Re}(\underline{Z}_{11}) = R_{\mathrm{E}} \cdot \frac{1}{1 + \frac{1}{\beta_{\mathrm{f}}} + R_{\mathrm{Bi}}\frac{C_{\mu}C_{\mathrm{BCx}}}{C_{\mu}+C_{\mathrm{BCx}}}} + \frac{R_{\mathrm{Bi}} + R_{\mathrm{Bx}}}{(1+\frac{1}{\beta_{\mathrm{f}}})(1+\frac{C_{\mu}}{C_{\mathrm{BCx}}}) + \beta_{\mathrm{f}}C_{\mu}R_{\mathrm{Bi}}},$$

$$(4.3.3\text{-}15)$$

which is close to (4.3.3-13) for sufficiently large values of the current amplification and using a lumped value for the base resistance.

4.3.3.3 H-parameter method

The H-parameter method [HS09] is based on a reformulation of the g_{mi} method. The internal transconductance is determined more precisely than in the original method by deembedding the equivalent circuit elements that are not of interest in multiple mathematical steps from the measured small-signal parameters. A similar approach was followed in [Pra92], but due to the more advanced EC used in [HS09], the latter is evaluated here.

The deembedding steps in the method do not rely on prior knowledge of extracted parameters, but instead on calculating the impact of external transistor elements on certain measured two-port parameters and removing them through multiple matrix operations. The employed EC considers an external base resistance in addition to the elements shown in fig. 4.8b.

The exact derivation is not repeated here. The emitter resistance is extracted according to

$$R_{\mathrm{E}} = min\left(\frac{\mathrm{Im}(\underline{\tilde{H}}_{11e})}{\mathrm{Im}(\underline{\tilde{H}}_{21e})} - \mathrm{Re}\left(\frac{1}{\underline{\tilde{H}}_{21e}}\right)\frac{m_{\mathrm{BEi}}V_{\mathrm{T}}}{I_{\mathrm{BEi}}}\right), \qquad (4.3.3\text{-}16)$$

where I_{BEi} is the internal base current and the nonideality factor of the internal base current m_{BEi} was added compared to the original publication, where it is assumed to be close to 1. The parameters $\underline{\tilde{H}}_{11e}$ and $\underline{\tilde{H}}_{21e}$ are the so-called unilateralized H-parameters calculated from the measured Y-parameters as

$$\underline{\tilde{H}}_{11e} = \frac{1}{\underline{Y}_{11} + \underline{Y}_{12}} \quad \text{and} \quad \underline{\tilde{H}}_{21e} = \frac{\underline{Y}_{21} - \underline{Y}_{12}}{\underline{Y}_{11} + \underline{Y}_{12}}. \qquad (4.3.3\text{-}17)$$

A variation of the method that both further deembeds parasitic elements of the transistor in analogy to the deembedding that is performed on measured S-parameters and takes into account the self-heating of the device was published in [RAH11] and is also evaluated here. The application of the method to simulated data is shown in fig. 4.11a.

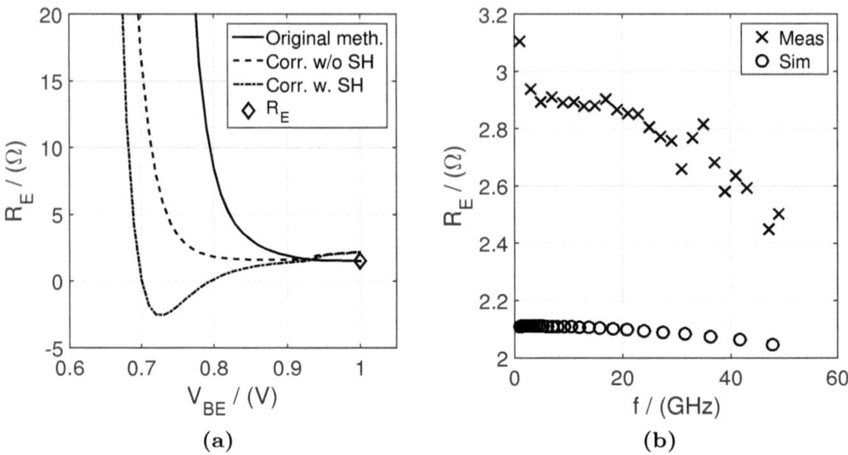

Figure 4.11: (a) Application of the H-parameter method to compact model generated data based on a GCS device with $A_{\mathrm{E0}} = 15x0.8$ µm^2. The extracted value for R_{E} is marked by the diamond. (b) Frequency dependence of the extracted value for measured and simulated data.

The main potential error source of the method are the assumptions made for its derivation in order to arrive at a sufficiently simple expression from a complicated equivalent circuit, e.g. the assumption of a purely internal base current. Additionally, the temperature dependence of the parameters is neglected. Attempting to take it into account according to [RAH11] causes an overcorrection, yielding negative values for the emitter resistance as demonstrated in fig. 4.11a. Additionally, the extraction is supposed to use the minimum of (4.3.3-16), which cannot be found for the investigated devices without risking thermal breakdown of the devices under test, so the lowest measurable value must be used for III-V HBTs. Finally, the frequency dependence of the main equation is not considered in the original publication, but can cause a significant deviation according to fig. 4.11b.

4.3.3.4 Ideal I_{B} method

The ideal I_{B} method [NT84] has been widely applied to SiGe devices in the past. It is based purely on a DC gummel measurement and therefore very easy to implement in a standard measurement setup. The method relies on the deviation of the base current from its ideal characteristic, extrapolated from low current densities, to determine the resistances. Since it assumes the base current to be affected by R_B and R_E, either can be extracted with it if the other is given. In this section, the application to R_E is discussed.

The ideal base-emitter current is given by

$$I_\mathrm{BE} = I_\mathrm{BEs}\left[\exp\left(\frac{V_\mathrm{B'E'}}{m_\mathrm{BE}V_\mathrm{T}}\right) - 1\right], \qquad (4.3.3\text{-}18)$$

where I_BEs is the BE saturation current and the -1 in the square brackets can be neglected if the internal BE voltage $V_\mathrm{B'E'}$ is much larger than $m_\mathrm{BE}V_\mathrm{T}$.

Rearranging the ideal base current equation and taking into account the impact of the series resistances using the simplified equivalent circuit in fig. 4.8a yields

$$\frac{m_\mathrm{BE}V_\mathrm{T}}{I_\mathrm{C}}\ln\left(\frac{I_\mathrm{B0}}{I_\mathrm{B}}\right) = \left(R_\mathrm{E} + \frac{R_\mathrm{Bi}}{B_\mathrm{f}}\right) + \frac{R_\mathrm{E} + R_\mathrm{Bx}}{B_\mathrm{f}}, \qquad (4.3.3\text{-}19)$$

with the forward DC current gain B_f, the ideal base current I_B0 and all other parameters as previously defined. The nonideality factor of the base current m_BE was added in (4.3.3-18) and (4.3.3-19) compared to [NT84].

The emitter resistance is extracted from a linear regression of the measurable r.h.s. of (4.3.3-19) w.r.t. $1/B_\mathrm{f}$. It is assumed that the bias-dependence of the internal base resistance and the current amplification cancel. Therefore, the constant part of the regression gives the sum $R_\mathrm{E} + R_\mathrm{Bi}/B_\mathrm{f}$ and the slope the term $(R_\mathrm{E}R_\mathrm{Bx})$.

The method suffers from various shortcomings. First, self-heating is neglected but occurs in the region with a measurable voltage drop across the series resistances. Second, an ideal I_B characteristic is assumed. This is a problem in InP HBTs since the base current is dominated by recombination rather than backinjection as has been discussed previously. Third, the bias dependence of R_Bi does not cancel with that of $1/B_\mathrm{f}$.

An application example of the method is shown in fig. 4.12a, while fig. 4.12b visualizes the slope of the logarithmic base current w.r.t. the base voltage. In the former figure, it is clearly visible that no proper extraction region exists and the slope of the r.h.s. of (4.3.3-19) is negative over a wide range. The reason becomes visible in the latter figure; if the assumption of an ideal base current were correct, the derivative would be

$$\frac{\mathrm{d}\ln(I_\mathrm{B}/\,\mathrm{A})}{\mathrm{d}V_\mathrm{BE}} = \frac{\mathrm{d}\ln\left(I_\mathrm{BEs}/\,\mathrm{A}\exp\left(\frac{V_\mathrm{BE}}{m_\mathrm{BE}V_\mathrm{T}}\right)\right)}{\mathrm{d}V_\mathrm{BE}} = \frac{1}{m_\mathrm{BE}V_\mathrm{T}}, \qquad (4.3.3\text{-}20)$$

and thus only voltage-dependent via self-heating. The impact of series resistances would decrease the slope. However, in fig. 4.12b, an increase of the slope is visible for most technologies between $V_\mathrm{BE} = 0.6\,\mathrm{V}$ and $V_\mathrm{BE} = 0.8\,\mathrm{V}$,

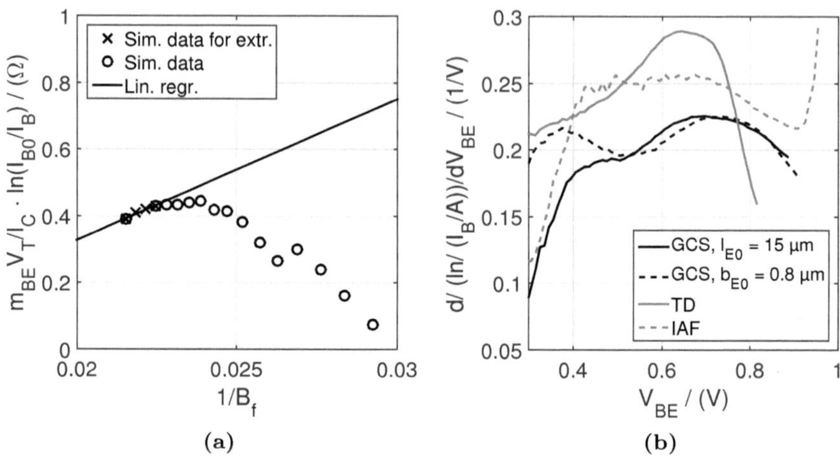

Figure 4.12: (a) Application of the ideal I_B method to compact model generated data based on a GCS device with $A_{E0} = 15x0.8\ \mu m^2$. (b) Logarithmic slope of the measured base current w.r.t. to V_{BE}. for one device each of all investigated technologies.

corresponding to the medium current region where it cannot be explained by self-heating. This non-ideal base current characteristic, coupled with the impact of self-heating in the high-current region, leads to the complete failure of the ideal I_B method.

4.3.3.5 Open collector method

The open collector method as applied here was derived in [GR98] and considers the measured CE voltage for an open collector node (i.e. $I_C = 0$) for the extraction of the emitter resistance. It is based on an in-depth investigation of the physical origin of the CE voltage under these circumstances. Since, in a III-V HBT, the substrate current is effectively 0 due to the semi-insulating substrate, the derived equation for the case $-I_n = I_p$, i.e. equilibrium of the electron and hole current, applies. The equation reads

$$V_{CE} = R_E I_E + \frac{2V_T}{1 + \frac{\mu_{n,c}}{\mu_{p,c}}} \ln(1 + \sqrt{I_E/I_{oS}}), \qquad (4.3.3\text{-}21)$$

with the collector electron and hole mobilities $\mu_{n,c}$ resp. $\mu_{p,c}$ and the auxiliary current I_{oS}. These three parameters can be determined from substrate transistor measurements; since those measurements are not possible for III-V

HTBs, (4.3.3-21) is instead simplified to

$$V_{\mathrm{CE}} = R_{\mathrm{E}} I_{\mathrm{E}} + a \ln(1 + \sqrt{I_{\mathrm{E}}/b}), \qquad (4.3.3\text{-}22)$$

in which the parameters R_{E}, a and b are determined by nonlinear curvefitting of the V_{CE} vs. I_{E} characteristic. An example of the application of the nonlinear fit to simulation data is shown in fig. 4.13.

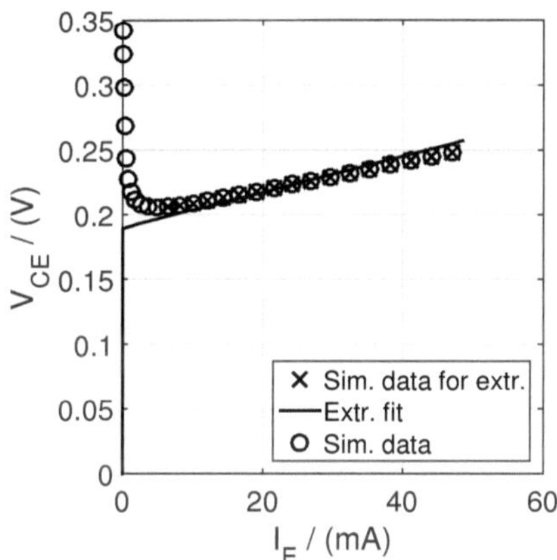

Figure 4.13: Application of the open collector method to compact model data (symbols) via interpolation (line) based on a GCS device with $A_{\mathrm{E0}} = 15x0.8\ \mu\mathrm{m}^2$. The emitter resistance is determined as one of the fit parameters.

The main error source for the method appears to be self-heating, which can become quite large towards the upper end of the extraction region. Additionally, for low current values, the CE voltage does not tend monotonously to zero, as expected from (4.3.3-21), but increases as the BC current, compensating the injected current, must fall to zero. The exact behavior depends on the ratio of BC and BE saturation currents and nonideality factors.

4.3.3.6 Simultaneous R_{E}-R_{th} method

In principle, the simultaneous R_{E}-R_{th} method, published in [TSW$^+$97], is very similar to the ideal I_{B} method in that it uses an extrapolated ideal base current according to (4.3.3-18) compared to the actual current to calculate the resistance value. However, it calculates and compensates for the impact of self-heating by simultaneously determining the thermal resistance from the previously extracted temperature dependence of the base current.

The implementation of the method consist of an iterative calculation of the two variables, executed until sufficient self-consistency is obtained, i.e. the change of either variable is sufficiently small. First, the emitter resistance is calculated from

$$I_\mathrm{E} R_\mathrm{E} = \Delta V_\mathrm{RE} = m_\mathrm{BE} V_\mathrm{T} \ln \left(\frac{I_\mathrm{B0}}{I_\mathrm{B}} \right) - I_\mathrm{B} R_\mathrm{B}, \qquad (4.3.3\text{-}23)$$

with the measured emitter current I_E and the isothermal currents I_B0, the ideal base current, and I_B, the measured, temperature-corrected base current. Then, using the temperature dependence of the base current

$$I_\mathrm{B} = I_\mathrm{B}(T_0) \left(\frac{T}{T_0} \right)^{\zeta_\mathrm{BET}} \exp \left[\frac{V_\mathrm{gEeff}}{V_\mathrm{T}} \left(\frac{T_\mathrm{j}}{T_0} - 1 \right) \right], \qquad (4.3.3\text{-}24)$$

with the temperature dependence parameters ζ_BET and V_gEeff as defined in HICUM (cf., e.g., [SC10]), the reference temperature T_0 and the junction temperature T_j, the thermal resistance is given by

$$R_\mathrm{th} = \frac{T_\mathrm{j}^2 k_\mathrm{B}/q}{V_\mathrm{gEeff} - V_\mathrm{B'E'}/m_\mathrm{BE} + \zeta_\mathrm{BET} T_\mathrm{j} k_\mathrm{B}/q - \alpha_\mathrm{Bf} T_\mathrm{j}^2 k_\mathrm{B}/q} \cdot \frac{\mathrm{d} \left(\ln \frac{I_\mathrm{B}(T_\mathrm{j})}{I_\mathrm{B}(T_0)} \right)}{\mathrm{d}T},$$
$$(4.3.3\text{-}25)$$

with the temperature coefficient of the current gain α_Bf. (4.3.3-25) requires the emitter resistance for the calculation of $V_\mathrm{B'E'}$. (4.3.3-24) and (4.3.3-25) were updated in [Kra15] compared to [TSW$^+$97] to include the more flexible HICUM temperature dependence. The isothermal base current is determined from (4.3.3-24).

An example of the extraction of R_th from the change of I_B is shown in fig. 4.14a and for the extraction of R_E based on ΔV_RE in fig. 4.14b.

Problems for the application of this method are caused by the non-ideal base current that cannot easily be described by a single diode equation (c.f. fig. 4.12b). However, if the extraction of the base current parameters is focused on the relevant region, the effect is at least partially compensated for, indicating that the major error source of the ideal I_B method is the self-heating of the transistor.

4.3.3.7 ΔT method

Like the simultaneous R_E-R_th method, the ΔT method [PLS14] concurrently extracts R_E and R_th, using the former to correctly calculate the internal voltages and the latter to calculate isothermal device properties.

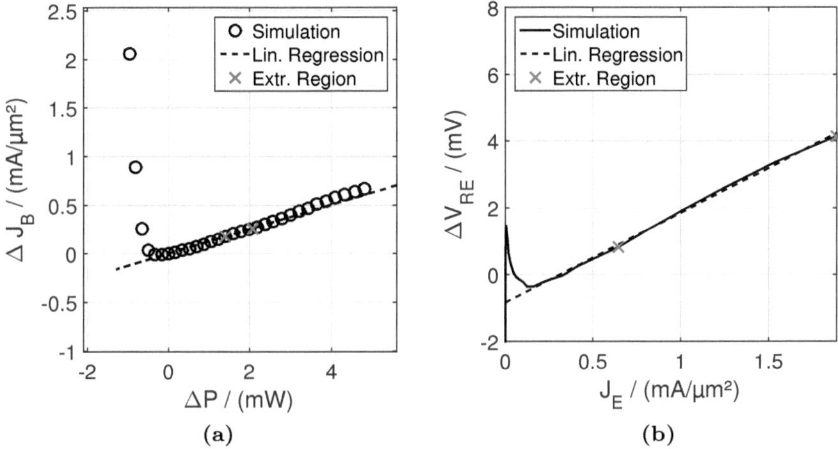

Figure 4.14: Extraction of (a) R_E and (b) R_th based on the simultaneous R_E-R_th method for compact model generated data based on a GCS device with $A_\mathrm{E0} = 15x0.8\ \mu\mathrm{m}^2$.

Starting from the ideal diode equation (4.3.3-18), the internal BE voltage is determined from the base current and from the applied voltage and the emitter resistance as

$$V_\mathrm{B'E'} = V_\mathrm{BE} - I_\mathrm{E} R_\mathrm{E}, \qquad (4.3.3\text{-}26)$$

where the base resistance was neglected due to the typically high current amplification in SiGe HBTs. Determining $V_\mathrm{B'E'}$ from the ideal base current only works for isothermal currents, since self-heating introduces too large an error otherwise. The difference of the device temperature compared to the ambient temperature is calculated from the collector current and the internal CE voltage, yielding

$$\Delta T = T - T_0 = [V_\mathrm{CE} - (I_\mathrm{E} R_\mathrm{E} + I_\mathrm{C} R_\mathrm{Cx})] I_\mathrm{C} R_\mathrm{th}, \qquad (4.3.3\text{-}27)$$

where ΔT is calculated from the known temperature dependence of the base current. R_E is then determined from a linear extrapolation of (4.3.3-27) w.r.t. V_CE for a constant, low I_C at

$$V_\mathrm{CE0} = V_\mathrm{CE}(\Delta T = 0) = I_\mathrm{E} R_\mathrm{E} + I_\mathrm{C} R_\mathrm{Cx}. \qquad (4.3.3\text{-}28)$$

Note, that finding ΔT from I_B requires a value for R_E to be known. Initally, it is assumed to be 0. During the extraction, R_E is calculated from

(4.3.3-26) and (4.3.3-28) until a self-consistent value is obtained. Finally, the thermal resistance is calculated from the linear regression of ΔT w.r.t. V_{CE}. An example for a single iteration step is shown in fig. 4.15.

Besides similar considerations regarding the applicability of an ideal base current diode equation for III-V HBTs as for the ideal I_{B} and simultaneous R_{E}-R_{th} methods, possible error sources for this method include neglecting the base resistance and the impact of the base current on self-heating, since the current amplification for InP HBTs is much lower than for SiGe HBTs.

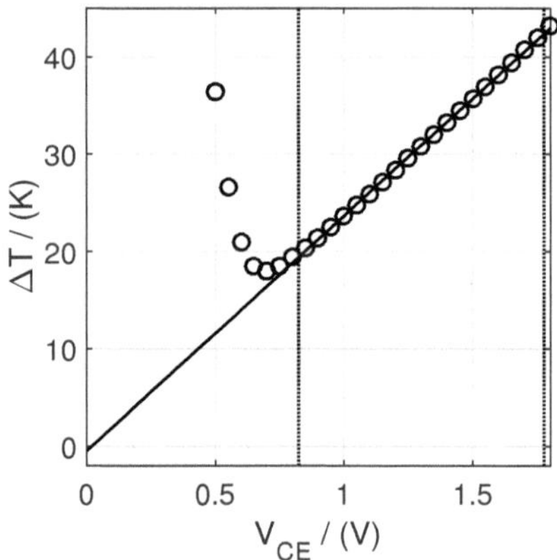

Figure 4.15: Example for the extraction of R_{E} from V_{CE0} for one iteration during the application of the ΔT method to compact model data (symbols) via linear extrapolation (line) based on a GCS device with $A_{\mathrm{E0}} = 15x0.8$ µm.

4.3.3.8 Results

The relative error of the emitter resistance obtained by applying the extraction methods to compact-model generated data with respect to the model card value for R_{E} is shown in tab. 4.2 while the absolute extracted values using the various methods are compared with the model card values in fig 4.16. The result of the method application to measured data can be found in fig. 4.17.

The main error sources for each method were already discussed in the method summaries. From a practical point of view, most of the methods presented here can easily be included in a standard measurement setup. The exception is the ΔT method, which requires a constant I_{C} for the extraction; however, from experience, this is quite reliably obtainable for InP HBTs during forced-I_{B} measurements and poses no major difficulty.

Table 4.2: Relative error (in %) of the emitter resistance for various extraction methods and different HBT sizes.

Tech	GCS, $l_{E0} = 15$ µm					$b_{E0} = 0.8$ µm			TD	IAF
A_{E0} / µm²	0.5x15	0.8x15	1x15	1.5x15	2x15	0.8x3	0.8x5	0.8x10	0.5x10	0.7x4
g_{mi}	40	51	59	78	92	23	25	24	-2	-50
ideal I_B	-35	-58	-78	-109	-127	-33	-18	-47	-8	-44
Open coll.	-24	-107	-154	-250	-295	-19	-29	12	-80	51
Z-param.	34	43	48	71	85	38	35	34	8	20
Z-param incl. corr.	31	34	35	34	33	30	28	27	8	20
H-param.	46	49	47	36	22	23	18	15	-2	-25
Simult.	-1	2	2	-8	-22	-16	-20	-26	-13	4
ΔT	17	32	43	63	74	23	34	41	-64	38

Application of the methods to measured data generally yields the same trend of the extracted compared to the compact model values. This verifies the use of compact-model generated data for this study. The main exception from this is the open collector method, which relies on reverse-bias characteristics that are commonly not important for circuit applications and were not modeled as precisely; the method still does not show the correct trend vs. geometry when applied to measurements. For the Z-parameter method, the correction was not applied to measured data since the required values were not known for all devices.

While the results of the ideal I_B method in tab. 4.2 do not show a large error for many devices, this is due to a very careful selection of the bias range - often only two bias points for the linear extrapolation in order to obtain a positive slope at all - and attributed to coincidence. The method is not recommended for practical use and not shown in the following figures.

Based on this study, two methods stand out. The first one is the simultaneous method, which gives the by far most accurate results when compared to the model and should be used when the base current temperature behavior is well known and conforms to the model. The second is the Z-Parameter method, which provides accurate values for R_E when a correction term is applied. Which method is most preferable depends also somewhat on the exact properties of the technology. Additionally, it has been shown that the open-collector method, which is still used for extracting R_E of III-V HBTs (e.g. [CSHB01]) does not yield reliable results at all and should not be employed.

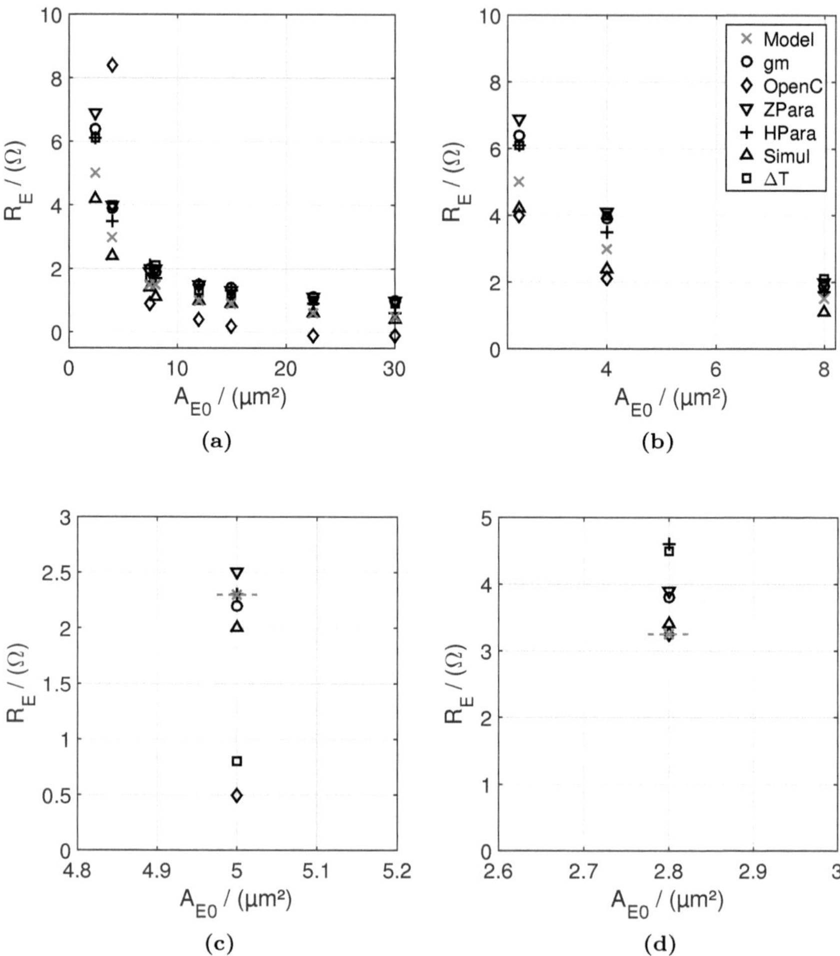

Figure 4.16: Results of the extraction methods applied to compact model data; (a) GCS $l_{E0} = 15$ µm, (b) GCS $b_{E0} = 0.8$ µm, (c) IAF and (d) Teledyne. The legend in (b) applies to all figures. The dashed line in (c) and (d) indicates the model value.

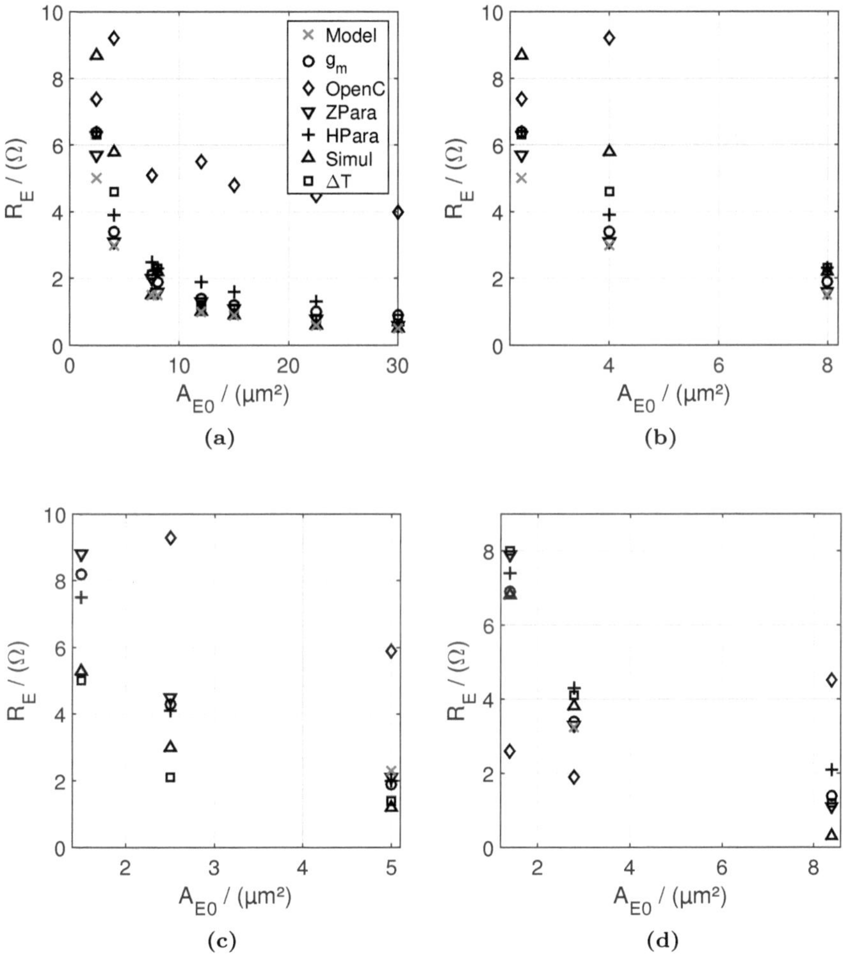

Figure 4.17: Results of the extraction methods applied to measured data.; (a) GCS $l_{E0} = 15$ µm, (b) GCS $b_{E0} = 0.8$ µm, (c) IAF and (d) Teledyne. The legend in (a) applies to all figures.

4.3.4 Methods for the collector resistance

The external collector resistance, while less important for the DC characteristics of a device than the emitter resistance, influences the RF behavior and
associated figures-of-merit such as the power gain. From a modeling point
of view, a badly chosen value for R_{Cx} often leads to wrong values for time
constants or maximum junction capacitance values.

Compared to, e.g., the emitter resistance, a relatively small number of extraction methods is available for R_{Cx} [MH82, VCT$^+$93, Gia72, Log71, GTB97,
Hus05], in part because the collector resistance can be accurately determined
from special test structures (TLMs, cf. section 4.5.2) and knowledge of the
device geometry. However, when no test structures are available or the exact
device geometry after fabrication is uncertain, or the frequency dependence
needs to be investigated, R_{Cx} has to be determined from measurements on
just a single transistor.

From the already limited number of extraction methods, those that rely on
the substrate transistor characteristics are unsuitable for III-V devices due to
the semi-insulating substrate. Furthermore, methods using impact ionization
are unsuitable for these technologies because of the very high risk of device
destruction when employing them [Zam16].

4.3.4.1 Open emitter method

The open emitter method [Gia72] exists as an analogy to the previously discussed open collector method (cf. section 4.3.3.5). However, the device physics
have been less thoroughly analyzed and the extraction is simply based on the
inverse slope of the linear regression on the measured $I_B(V_{EC})$ so that

$$V_{EC}\big|_{I_E=0} \approx R_{Cx}I_B. \tag{4.3.4-29}$$

An example is given in fig. 4.18. The extraction range for the method is
easy to identify and the exact chosen values for its limits have little impact
on the extracted value. No prior extraction steps are needed. While SiGe
devices in a standard GSG pad layout for RF measurements cannot be used
for this extraction method due to the shorted emitter-substrate contacts, this
is irrelevant for III-V devices because of the semi-insulating substrate, though
a separate DC measurement still needs to be set up for the floating emitter
node.

A possible error sources for the method, apart from self-heating for higher
values of I_B, is the internal EC voltage $V_{E'C'}$, which is assumed here as independent of the base current.

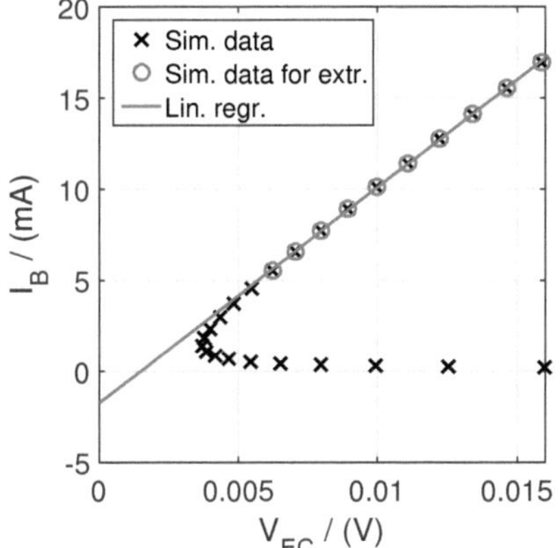

Figure 4.18: Example for the extraction of R_{Cx} using the open emitter method based on compact model data (symbols) via linear regression (line) for a GCS device with $A_{\mathrm{E0}} = 15x0.8\ \mathrm{\mu m}^2$.

4.3.4.2 Forced B_{f} method

Similar to the open emitter method, the forced B_{f} method [VCT$^+$93] assumes that, for a given DC current gain $B_{\mathrm{f}} < B_{\mathrm{f0}}$ with B_{f0} as the low-injection forward current gain, the change of the CE voltage is related only to the change of the collector current and the series resistances, i.e. the internal CE voltage remains constant. The voltage change is then described as

$$\Delta V_{\mathrm{CE}} = \Delta I_{\mathrm{C}} \left[R_{\mathrm{Cx}} + \left(1 + \frac{1}{B_{\mathrm{f}}}\right) R_{\mathrm{E}} \right]. \tag{4.3.4-30}$$

With a known value for the emitter resistance R_{E} and a constant, measurable B_{f}, the collector resistance can be determined from the slope of V_{CE} w.r.t. I_{C}. An example is shown in fig. 4.19, which also plots the internal CE voltage $V_{\mathrm{C'E'}}$ taken directly from the model data during simulation. $V_{\mathrm{C'E'}}$ is clearly not constant in the relevant range in fig. 4.19.

In order to improve the extraction method, an attempt is made to include the current dependence of $V_{\mathrm{C'E'}}$ using the previously extracted base current parameters and the equation for the total base current

$$I_{\mathrm{B}} = I_{\mathrm{BEs}} \exp\left(\frac{V_{\mathrm{B'E'}}}{m_{\mathrm{BE}} V_{\mathrm{T}}}\right) + I_{\mathrm{BCs}} \exp\left(\frac{V_{\mathrm{B'C'}}}{m_{\mathrm{BC}} V_{\mathrm{T}}}\right). \tag{4.3.4-31}$$

Rearranging (4.3.4-31) and keeping in mind that $V_{\mathrm{C'E'}} = V_{\mathrm{B'E'}} - V_{\mathrm{B'C'}}$ yields

$$V_{\mathrm{C'E'},\mathrm{corr}} = -m_{\mathrm{BC}}V_{\mathrm{T}}\ln\left[\frac{I_{\mathrm{B}} - I_{\mathrm{BEs}}\exp\left(\frac{V_{\mathrm{B'E'}}}{m_{\mathrm{BE}}V_{\mathrm{T}}}\right)}{I_{\mathrm{BCs}}\exp\left(\frac{V_{\mathrm{B'E'}}}{m_{\mathrm{BC}}V_{\mathrm{T}}}\right)}\right] \tag{4.3.4-32}$$

as the correction voltage.

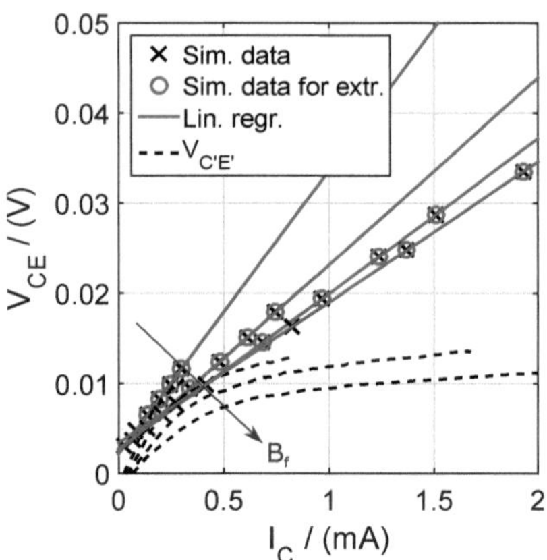

Figure 4.19: Example for the extraction of R_{Cx} using the forced B_{f} method based on compact model data (symbols) via linear regression (line) for an IAF device with $A_{\mathrm{E0}} = 4x0.7\ \mu\mathrm{m}^2$ for multiple values of the forced current gain. The dashed lines indicate the internal CE voltage.

Even including the correction, the method suffers from the inaccurate determination of $V_{\mathrm{C'E'}}$ due to the simplified base current equation and the lack of an precise determination of $V_{\mathrm{B'E'}}$, for which the base and emitter resistances must be known with good accuracy. Additionally, in a realistic measurement setup, forcing two input currents is quite difficult; instead, for a given I_{B}, V_{BE} must be swept to obtain the desired B_{f}, requiring a high number of measurement points.

4.3.4.3 Z-parameter method

The principle of the Z-parameter [MT92, GTB97] method when applied to the collector resistance is similar to the application to R_{E} (c.f. section 4.3.3.2): the evolution of the real part of a Z-parameter is extrapolated from the medium to high current range w.r.t. $1/I_{\mathrm{B}}$ towards infinite currents, i.e. $1/I_{\mathrm{B}} = 0$, assuming that the diode conductances and the transconductance are then infinite, the internal transistor is shorted and the equivalent circuit consists only of the series resistances. Given these circumstances,the collector resistance is determined as the intersection of the extrapolation of $\mathrm{Re}(\underline{Z}_{22} - \underline{Z}_{21})$ with the y-axis.

Note, that for application to the collector resistance, the transistor must be in saturation, though this is not spelled out explicitly in either [MT92] or [GTB97], and the extrapolation w.r.t. $1/I_\mathrm{B} = 0$ is correct in this case.

As discussed previously, error sources for this method include the insufficiently accurate equivalent circuit, the impact of self-heating and the chosen extraction frequency. The latter, however, appears to have a minor impact for most technologies, as shown in fig. 4.20a. The only exception is the GCS technology with $l_\mathrm{E0} = 15\mu\mathrm{m}$ due to a high BC leakage current in this process run, which does not exist for the transistors with $b_\mathrm{E0} = 0.8\mu\mathrm{m}$ (not shown in the figure for reasons of clarity) While this effect is unusual for III-V technologies, it still demonstrates the need to verify extraction method results.

An example for the application of the Z-parameter method for the collector resistance to simulation data is shown in fig. 4.20b.

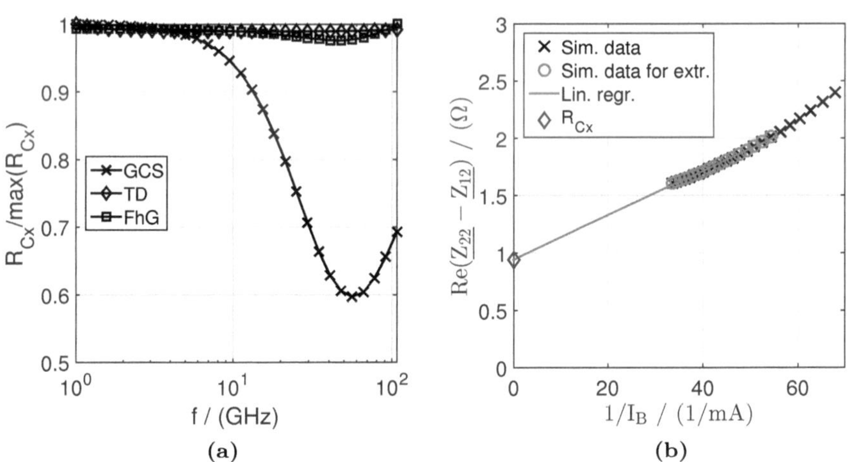

Figure 4.20: (a) Frequency dependence of the extraction of R_Cx using the Z-parameter method the GCS devices $A_\mathrm{E0} = 0.8x15\ \mu\mathrm{m}^2$ and the modeled TD and IAF devices normalized to the extracted maximum value and (b) example of the application of the Z-parameter method to compact model data (symbols) for a GCS device with $A_\mathrm{E0} = 15x0.8\ \mu\mathrm{m}^2$.

4.3.4.4 H-parameter method

The H-parameter method for the extraction of the collector resistance [Hus05] relies on the same advanced equivalent circuit as the H-parameter method for the emitter resistance (cf. section 4.3.3.3). No extraction range is specified, though from the extraction examples, it can be concluded that small-signal measurements in the forward active transistor region are used. On this basis, an equation for R_{Cx} is developed that reads

$$R_{\mathrm{Cx}} = -\frac{\frac{1}{\omega} \cdot \mathrm{Im}\left(\frac{1}{\tilde{H}_{21}}\right) + \left(\frac{1}{\omega_{\mathrm{T}}} + r_{\mathrm{Ci}} C_{\mathrm{jCi}}\right)}{\frac{1}{\omega} \cdot \mathrm{Im}\left(\frac{1}{\tilde{Z}_{21}}\right)}. \tag{4.3.4-33}$$

Here, $\tilde{H}_{21}$ is the unilateral h-parameter as defined in section 4.3.3.3; $\tilde{Z}_{21} = Z_{21} - Z_{12}$ is the analogously defined unilateral Z-parameter. ω_{T} is the pole of the internal current gain and r_{Ci} is the internal collector resistance (which is included in the transfer current formulation in HICUM). The remaining symbols have their usual meanings.

(4.3.4-33) contains too many unknowns to determine R_{Cx} directly. In order to resolve this, the term $\left(\frac{1}{\omega_{\mathrm{T}}} + r_{\mathrm{Ci}} C_{\mathrm{jCi}}\right)$ is calculated from the linear extrapolation of

$$y_{\mathrm{RCx}} = \frac{\mathrm{Im}\left[\frac{1}{\tilde{H}_{21}} \cdot \left(\frac{1}{\tilde{Z}_{21}}\right)^{*} - \frac{1}{\tilde{Z}_{21}}\right]}{\omega \cdot \mathrm{Re}\left(\frac{1}{\tilde{Z}_{21}}\right)} \tag{4.3.4-34}$$

with respect to

$$x_{RCx} = \frac{\omega \cdot \mathrm{Im}\left(\frac{1}{\tilde{Z}_{21}}\right)}{\mathrm{Re}\left(\frac{1}{\tilde{Z}_{21}}\right)}, \tag{4.3.4-35}$$

with the transposed inverse unilateral Z-parameter $\left(\frac{1}{\tilde{Z}_{21}}\right)^{*}$, as shown in the example in fig. 4.21a.

The extraction method makes some unjustified assumptions on the model; while the original publication indicated good agreement with expected values, for III-V devices, the obtained values for R_{Cx} are strongly bias dependent and physically meaningless, as shown in fig. 4.21b. Reasonable results could not be obtained for any of the investigated InP technologies.

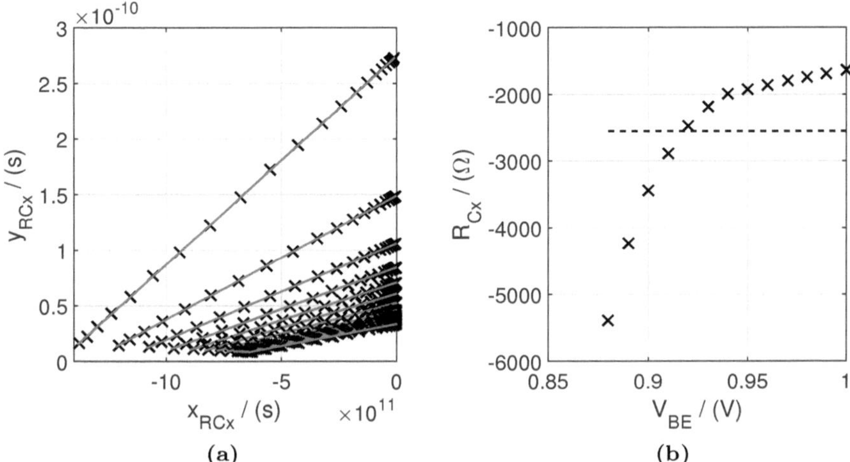

Figure 4.21: (a) Extrapolation of the unknown term in (4.3.4-33) from (4.3.4-34) w.r.t. (4.3.4-35) (lines show the extrapolation, symbols the model results) and (b) result of the application of the H-parameter method to compact model data (symbols), both for a GCS device with $A_{E0} = 15x0.8\ \mu m^2$ for various bias points. The dashed line shows the extracted average R_{Cx}.

4.3.4.5 Results

The relative error of the R_{Cx} extraction w.r.t. the model card value is listed in tab. 4.3 while the absolute extracted value is shown in fig. 4.22 for for applying the method to the compact-model generated data and fig. 4.23 for application to measured data.

Based on the application of the methods to the data generated by compact model simulations, the H-Parameter method fails completely for InP HBTs. Exact reasons are currently unknown, and given that both the Z-parameter and the open emitter methods yield very good results with low average errors of 8% and 2%, respectively, a further investigation is not warranted. The correction for the internal CE voltage introduced in section 4.3.4.2 for the forced B_f method reduces the average error by nearly 40 percentage points to 87 %, but even then, the average accuracy is too low to recommend its use.

From a practical point of view, considering the two methods that extract R_{Cx} quite precisely, the Z-parameter method requires the lower effort, since the required measurement can be taken with standard GSG probes and the quite small GSG pads are difficult to use with DC probes, especially since a force-sense measurement would be required to remove cable and contact resistances. However, if a DC setup is available, it is recommended that both methods are used and the results are compared for added reliability.

Comparing the results of the method application to simulated respectively measured data, similar trends are observed in that the Z-parameter and the open emitter methods are closest to the expected result. Note, that a certain difference between the actual and modeled external collector resistance is possible and that the extraction methods applied to measured data may be closer to the real R_{Cx} than the model card value.

The model card value for the GCS technology with $l_{\mathrm{E0}} = 15\,\mu\mathrm{m}$ was determined using special test structures (cf. section 4.5.2). From the drawn geometry and known sheet resistance values, R_{Cx} should increase with increasing b_{E0}, as indicated by the model card data. Both the Z-parameter and the open emitter method, however, show decreasing R_{Cx} values with increasing emitter width. Since both methods agree, this could indicate non-standard scaling caused by, e.g., unknown actual (vs. layout) dimensions. This effect cannot be discovered using the test structure method unless additional information beyond electrical measurements is available, because standard scaling is inherently assumed in the calculations. The required verification of actual dimensions is a disadvantage of the test structure method. SEM/TEM pictures would be useful to draw final conclusions here, but were not available at the time of this writing for all devices.

Table 4.3: Relative error (in %) of the collector resistance for various extraction methods applied to model data and different HBT sizes.

Tech	GCS, $l_{E0} = 15\,\mu m$					$b_{E0} = 0.8\,\mu m$			TD	IAF
$A_{E0}\,/\,\mu m^2$	0.5x15	0.8x15	1x15	1.5x15	2x15	0.8x3	0.8x5	0.8x10	0.5x10	0.7x4
Open Emit.	-2	-2	-2	-2	-2	-2	-2	-2	-4	-4
Forced B_f	-150	-90	-67	-35	-22	167	104	136	-45	54
Z-param.	-14	2	-17	14	23	0	0	0	1	-9
H-param. $(/10^5)$	7	3	2	2	2	100	80	60	0.05	20

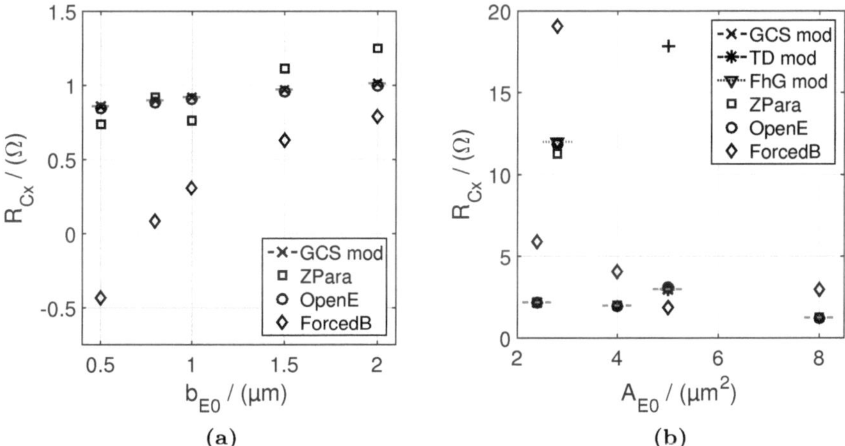

Figure 4.22: Results of the extraction methods applied to measured data.; (a) GCS $l_{E0} = 15\,\mu m$, (b) other technologies and sizes. The results of the H-parameter method are not shown due to the scale of the figure.

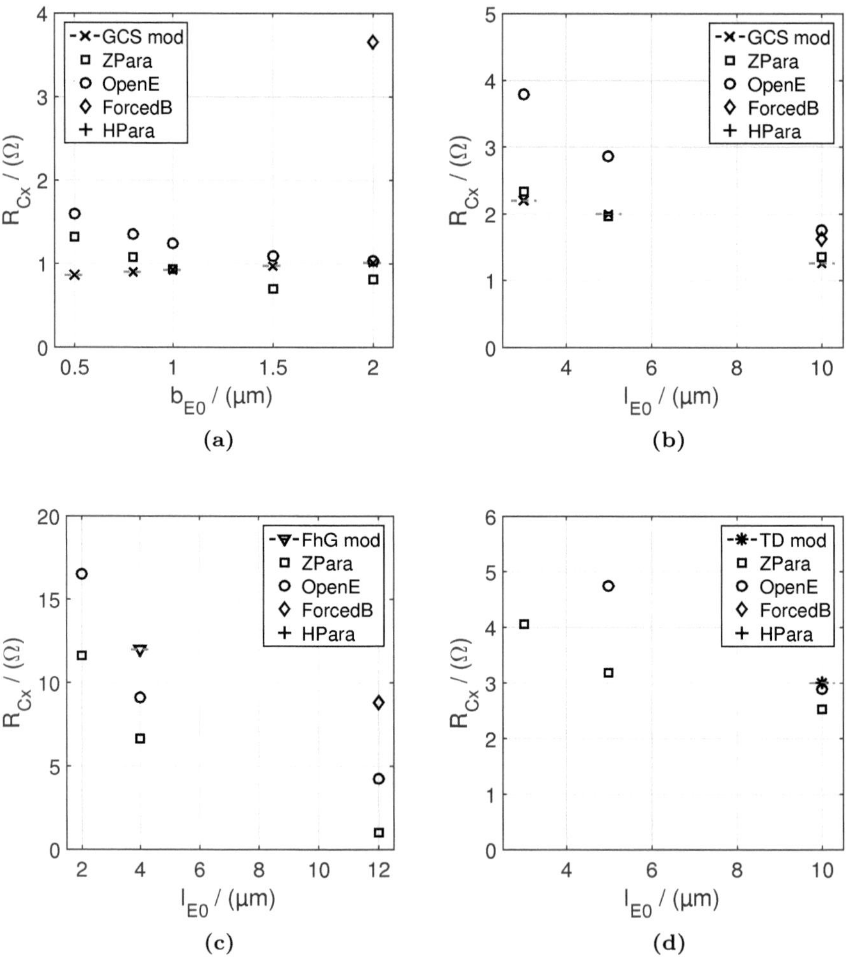

Figure 4.23: Results of the extraction methods applied to measured data.; (a) GCS $l_{E0} = 15\,\mu m$, (b) GCS $b_{E0} = 0.8\,\mu m$, (c) IAF and (d) Teledyne.

4.3.5 Methods for the base resistance

The base series resistance R_B impacts the high-frequency behavior of HBTs as well as its DC characteristics due to the voltage drop caused by the base current. This is especially true for InP HBTs due to the much lower DC current gain compared to, e.g., SiGe HBTs. The impact of R_B is expected to increase with advancing technology nodes due to shrinking vertical dimensions and correspondingly higher contact and partially also sheet resistances [RLB08, SRC+16]. A large number of methods exist in the literature for the determination of R_B based on terminal data (e.g. [VDK+96, Log71, ZBB+96, VCT+93, NT84, ZMCD91, MT92, GTB97, Pra92, BDE69, SM72, NH91, KPK99, KU75, LGG+98, McA06, Sch88, Neu87]), with, unfortunately, widely varying results when applied to the same transistor (e.g. [Kra15, VDK+96]).

An alternative to extraction methods based directly on terminal characteristics of a single transistor is the use of special test structures combined with knowledge about the transistor geometry. Such structures, including the tetrode [RS91, SL07], have been used for SiGe HBTs since at least the seventies, and the geometry dependence of the base resistance has been well known [Sch91a, SKLC08, YZ13]. They are discussed in this work in section 4.5.2.

As discussed in previous sections, methods based on the parasitic substrate transistor (e.g. [ZBB+96]) will not be investigated due to the semi-insulating substrate of III-V HBTs. Extraction of R_B from noise measurements, as originally described in [KU75], yielded reasonable agreement when applied to advanced SiGe HBTs [Kra15]. Similar results are expected for the noise-based method proposed in [LGG+98]. However, the overall effort for noise-based methods is usually too time consuming in industrial practice. Additionally, at high frequencies the impact of noise correlation and at high bias the effect of self-heating add uncertainty to the extraction results. A more simple method based on the change of the BE voltage with the base current was proposed in [McA06]. When applied in this work, it yielded only negative values even when the conductances were obtained from small-signal measurements instead of finite differences [KS15]. A possible error source is the very small output conductance of InP HBTs resulting from high base and low collector doping, a limitation that was also mentioned in [McA06]. In addition, the method relies on a forward $I_B(V_{BE})$ characteristic with an ideal portion, which is often difficult to find in III-V HBTs (cf. discussion in section 4.3.3.4). Finally, a low-frequency method was proposed in [Neu87], using a (special) bridge for measurement. The method is limited to low injection and and requires several other parameters to be known.

Since it is difficult to find an extraction method for the base resistance that consistently performs well, the investigation in this section was extended to include simulation data with deactivated self-heating, representing an ideal pulsed measurement. Since thermal breakdown is the main limitation of the measurement range, this also extends the maximum permissible current density. While measured data for these conditions are not available (cf. discussion in section 4.2.3), an investigation of the extraction method performance in principle is possible.

4.3.5.1 Ideal I_B method

The ideal I_B method was already discussed in section 4.3.3.4 for the application to the emitter resistance. If R_E is known from another source, the method can also be applied to the base resistance. Then, the internal, bias-dependent part of the base resistance R_{Bi} is determined from the constant part of the linear regression in (4.3.3-19) while R_{Bx} is determined from the slope.

Independent of the method application to R_E or R_B, the method suffers from two major shortcomings. First, self-heating is neglected but occurs in the region with a measurable voltage drop across the series resistances. Second, an ideal I_B characteristic is assumed. This is a problem in InP HBTs since the base current is dominated by recombination rather than backinjection [RPHJ06], and the nonideality factor of the diode current is not constant.

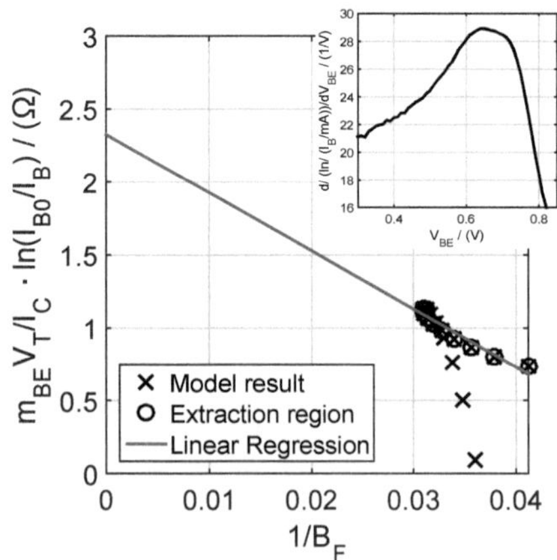

Figure 4.24: Application of the ideal I_B method: l.h.s. of (4.24) w.r.t. $1/B_f$ and linear regression for R_B extraction for a simulated TD device with $A_{E0} = 10x0.5$ μm.

The combination of these effects makes the method very unreliable for InP HBTs. Fig. 4.24 shows the linear regression used for the extraction from (4.24), exhibiting a negative slope and thus a negative extracted R_{Bx} as an example. The slope of $\ln(I_{\mathrm{B}})$ with V_{BE}, shown in the inset, clearly demonstrates non-ideal behavior for the entire bias range.

4.3.5.2 Open collector method

The open-collector method was originally proposed for determining R_{E} as discussed in section 4.3.3.5, but later extended in [ZMCD91] towards also extracting R_{B}. Since the collector is open ($I_{\mathrm{C}} = 0$), the base current equals the emitter current ($I_{\mathrm{B}} = I_{\mathrm{E}}$). Then, the resistance can be determined from

$$\frac{V_{\mathrm{BE}} - V_{\mathrm{CEs}}}{I_{\mathrm{B}}} = \frac{1}{I_{\mathrm{B}}}\left(V_{\mathrm{B'E'}} - V_{\mathrm{C'E's}}\right) + R_{\mathrm{B}}, \qquad (4.3.5\text{-}36)$$

with $V_{\mathrm{C'E's}}$ as the internal CE saturation voltage. The bias dependence of the term in the parentheses on the r.h.s. in the range relevant for the base resistance extraction is negligible, which has been verified by simulations. A linear regression of the measurable l.h.s. w.r.t. $1/I_{\mathrm{B}}$ can then be used to determine the base resistance as the intercept with the y-axis. For the technologies investigated here, the extraction region choice does not impact the result over a wide range of I_{B}.

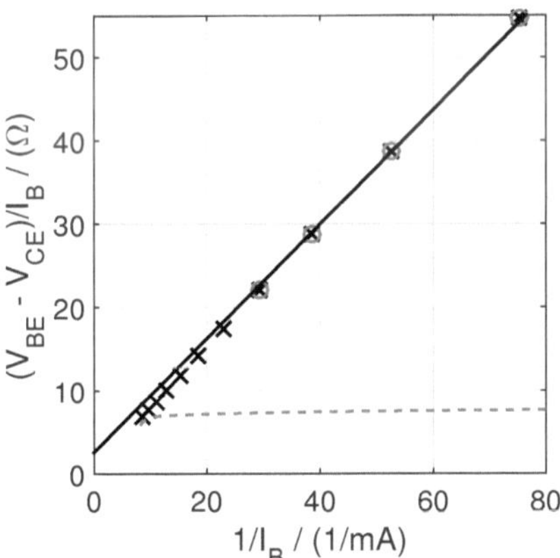

Figure 4.25: Application of the open collector method: l.h.s. of (4.3.5-36) (x), extraction range (o), linear regression (solid line) and bias-dependent total base resistance (dashed line) for a simulated TD device with $A_{\mathrm{E0}} = 0.5x10\ \mu\mathrm{m}^2$.

The method as applied to the base resistance neglects an external BC current flow, which is present in III-V mesa HBTs. This current does not flow through R_{Bi} and only partially through R_{Bx}, since the BC diode is located not only under the emitter, but extends all the way under the base contact (e.g. fig. 2.3). This causes the method to underestimate the base resistance (cf. fig. 4.25). Thus, the method is quite inaccurate for realistic devices.

A correction for the external base-collector current is possible, but requires width-scaled devices, which are not commonly available for InP (and other III-V) HBT processes, for separating internal and external currents.

4.3.5.3 Z-parameter method

The Z-parameter method as published in [MT92] and more explicitly described in [GTB97] was already discussed for the application to R_{E} in section 4.3.3.2 and R_{Cx} in section 4.3.4.3. The same principle can also be applied to the base resistance. No explicit bias conditions for the extraction are given in [MT92, GTB97], though recently [KS15], it was found that the method works best for SiGe devices when applied in saturation; then, the transistor is biased so that the internal conductances are nearly infinite and the nodes B', C' and E' can be assumed to be short-circuited. For the required high bias region, one obtains from the equivalent circuit in fig. 4.8b

$$\mathrm{Re}(\underline{Z}_{11} - \underline{Z}_{12}) = \frac{R_{\mathrm{B}}}{1 + \omega^2 C_{\mathrm{BCx}}^2 R_{\mathrm{B}}^2}, \qquad (4.3.5\text{-}37)$$

which is equal to R_{B} only if C_{BCx} is neglected or for sufficiently low frequencies. For InP HBTs operated in saturation, the very large external BC diode bridges not only R_{Bi}, but also parts of R_{Bx} due to the BC mesa extension all the way under the base contact. Therefore, the method must be applied under forward active conditions. Corresponding tests based on model data have verified this conclusion.

Fig. 4.26 shows an application example of the method both with and without self-heating for different frequencies. In the former case (cf. fig. 4.26a), the value of $\mathrm{Re}(\underline{Z}_{11} - \underline{Z}_{12})$ rises towards the high bias region, in contradiction to [GTB97], making the extraction region ambiguous. Without self-heating (cf fig. 4.26b), a wider bias range can be used since thermal breakdown is not an issue. From comparison with the model value, it is clear that the relative maximum of $\mathrm{Re}(\underline{Z}_{11} - \underline{Z}_{12})$ in the high-bias region agrees well with the total base resistance. The extraction frequency has little impact on the extracted resistance. Since the frequency should not be too low due to the required short-circuit approximations and not too high due to (4.3.5-37), $f_0 = 10\,\mathrm{GHz}$ was chosen here.

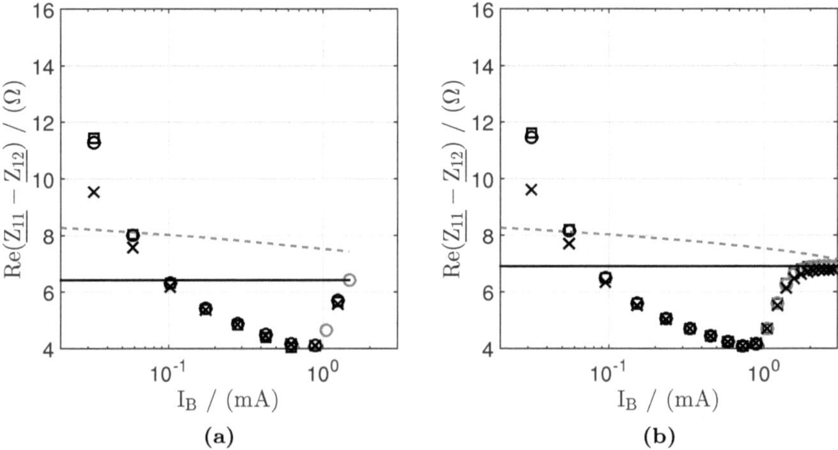

Figure 4.26: Application of the Z-parameter method to the compact model for TD $A_{\mathrm{E0}} = 0.5x10\ \mathrm{\mu m}^2$. (a) Full model with self-heating and (b) isothermal simulation: $\mathrm{Re}(\underline{Z}_{11}$ - $\underline{Z}_{12})$ from simulation for f $= 1\,\mathrm{GHz}$ (square), f $= 10\,\mathrm{GHz}$ (o) and f $= 40\,\mathrm{GHz}$ (x). The solid line represents the extracted R_{Bx} and the dashed line $R_{\mathrm{B}} = R_{\mathrm{Bx}} + R_{\mathrm{Bi}}$ of the compact model as the reference.

4.3.5.4 H-parameter method

In [Pra92], an extraction of the base resistance from the H-Parameters based on the simple hybrid-π equivalent circuit in fig. 4.8b without the external BC capacitance C_{BCx} was proposed. The resistance is extracted according to

$$R_{\mathrm{B}} = \mathrm{Re}(\underline{H}_{11}) - R_{\mathrm{E}} - \frac{r_\pi + \beta_{\mathrm{f0}} R_{\mathrm{E}}}{1 + x^2} \tag{4.3.5-38}$$

with $r_\pi = 1/g_\pi$, , $x = -\mathrm{Im}(\underline{\beta_{\mathrm{f}}})/\mathrm{Re}(\underline{\beta_{\mathrm{f}}})$ and $\underline{\beta_{\mathrm{f}}}$ as the forward current gain in common-emitter configuration with a low-frequency value of β_{f0}. Using a frequency close to the $3\,\mathrm{dB}$ corner frequency of the small-signal current gain, $f_\beta = \omega_\beta/2\pi$, is recommended in [Pra92]. This recommendation has been verified; smaller and larger frequencies both yield worse accuracy than f_β for medium and high bias. In [Pra92], r_π and β_{f0} are further approximated by $r_\pi = \beta_{\mathrm{f0}} V_{\mathrm{T}}/I_{\mathrm{C}}$ and $\beta_{\mathrm{f0}} = \mathrm{Re}(\underline{H}_{21}) \cdot (1 + x^2)$. In practice, the approximation of r_π introduces a large error in the extraction. However, by rearranging the equations in [Pra92], one obtains

$$R_\mathrm{B} = \mathrm{Re}(\underline{H}_{11}) - R_\mathrm{E} - \frac{\mathrm{Im}(\underline{H}_{11})}{1 + x^2}, \qquad (4.3.5\text{-}39)$$

which does not require the aforementioned approximations and has been used here.

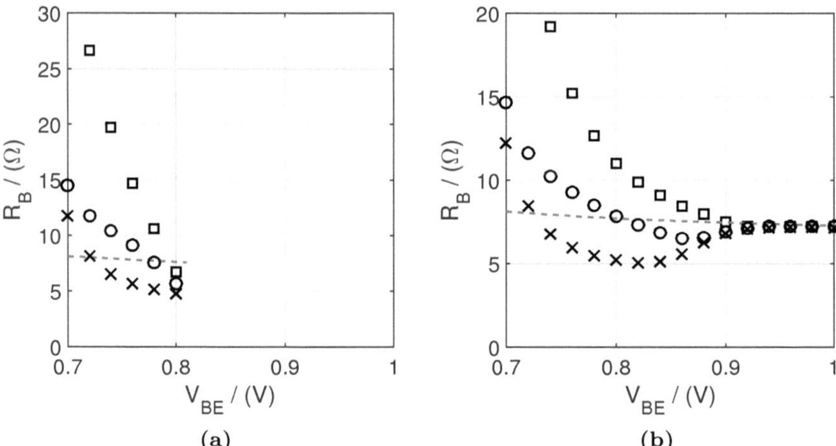

Figure 4.27: Application of the H-Parameter method to the compact model for TD $A_\mathrm{E0} = 0.5x10\,\mathrm{\mu m}^2$: (a) with and (b) without self-heating. Symbols represent extracted values at $f = 1\,\mathrm{GHz}$ (square), $f = f_\beta = 10\,\mathrm{GHz}$ (o) and $f = 40\,\mathrm{GHz}$ (x). The dashed line shows the actual $R_B = R_{Bx} + R_{Bi}$.

Fig. 4.27 shows an application example both with and without self-heating. From fig. 4.27a, it is evident that the extraction result is strongly bias-dependent and close to the target value only for very high bias. In fig. 4.27b, better agreement is obtained; this is only possible in isothermal simulations or with pulsed measurements, since the device would otherwise suffer thermal breakdown under these conditions.

At low bias, the main error source for this method is the too strong simplification of the equivalent circuit in the realistic bias range. Additionally, parasitic capacitances were not taken into account in the equivalent circuit and above formulations and may contribute to additional deviations.

4.3.5.5 Circle impedance method

This method is one of the oldest and most employed methods for determining the base resistance, having been originally proposed in [BDE69] with variants and refinements described in [SM72, NH91, KPK99]. In contrast to the other methods discussed so far, the frequency dependence is used to determine the base resistance. It is assumed that, for medium frequencies, the common-emitter input impedance $\underline{H}_{11}$ forms a semicircle in the complex plane. Extrapolating to infinite frequencies then gives the base resistance as intercept with the x-axis.

Compared to [BDE69], the method was extended in [NH91] to include the external BC capacitance by replacing $\underline{H}_{11}$ with

$$\underline{H}_{11,corr} = \frac{1}{\underline{Y}_{11} + \underline{Y}_{12}} = \frac{1 + R_{\mathrm{B}} + r_\pi + j\omega(C_\pi + C_\mu)}{g_\pi + j\omega C_\pi}. \tag{4.3.5-40}$$

In [KPK99] the emitter resistance is additionally taken into account, yielding

$$R_{\mathrm{B}} = \frac{H_{11,corr} - R_{\mathrm{E}}}{1 + \frac{C_\mu}{C_\pi}(1 + R_{\mathrm{E}}(g_{\mathrm{m}} + g_\pi))}. \tag{4.3.5-41}$$

In analyzing the error sources for this method, the parasitic capacitances present in the model but not considered in the simplified equivalent circuit were identified as a problem. In practice, their influence can be deembedded if they are determined from suitable test structures and geometry scaling. Additionally, it was found that (4.3.5-41) may not be sufficiently precise in correcting for the remaining equivalent circuit elements at high frequencies. A more complete analysis, neglecting second-order frequency dependent effects, yields

$$R_B = \frac{(1 + \frac{C_\pi}{C_\mu})H_{11,corr} - \frac{C_\pi}{C_\mu}R_E}{1 + \frac{C_\pi}{C_\mu} + R_E(g_m + g_\pi)}, \tag{4.3.5-42}$$

which exhibits better agreement.

Fig. 4.28a shows typical simulation data for a single bias point and multiple frequencies and the extrapolation based on a semicircle equation. Fig. 4.28b shows the extraction result over bias. For high bias, the extraction result comes close to the actual model value. However, since the target value is unknown in practice, it is difficult to select the bias range that yields physically accurate values. It is also clearly visible that for medium and high bias (4.3.5-42) performs somewhat better than (4.3.5-40) and (4.3.5-41).

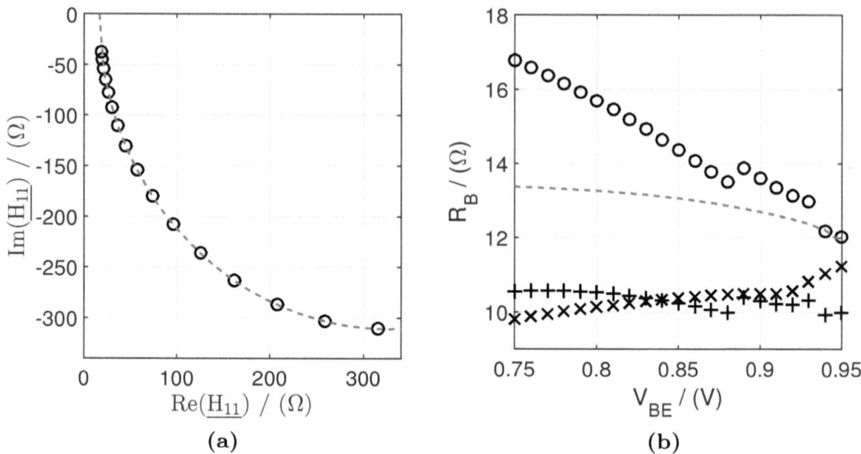

Figure 4.28: Application of circle impedance method for GCS $A_{E0} = 0.8x15$ µm^2. (a) Simulation result (o) with semicircle extrapolation (dashed line). (b) Bias-dependent extraction result according to (4.3.5-40) (x), (4.3.5-41) (+), and (4.3.5-42) (o) with the actual value for $R_B = R_{Bx} + R_{Bi}$ (dashed line) as reference.

4.3.5.6 Results

Compared to the evaluation for the emitter and collector resistances, the results of the investigation into the extraction methods for the base resistance are more difficult to analyze. The base resistance is split into the internal and external contributions; it is therefore of interest to determine whether both contributions can be determined by a method. Furthermore, in order to calculate the relative error of an extraction method w.r.t. the compact model value, the *bias-dependent* reference value must be used.

While for SiGe HBTs, R_{Bi} becomes small compared to R_{Bx} at high bias, this is not true for InP HBTs due the more limited measurement range (to avoid device destruction) and the much weaker bias dependence of R_{Bi}. Therefore, the extraction method results contain all of R_{Bx} and an unknown fraction of R_{Bi}. Using this value as a bias-independent R_B in a compact model would yield wrong results and a separation of the components must be considered here, even if it is not considered in the extraction methods.

The Z-parameter, open collector and ideal-I_B methods all give a single R_B value without a bias dependence. Therefore, R_{Bx} can only be determined from geometry scaling vs. the emitter width b_{E0}. This has been shown in fig.

4.29.

The H-parameter and circle impedance methods yield an extracted R_B for every bias point. Theoretically, the zero-bias internal base resistance R_{Bi0} and R_{Bx} can be determined from a nonlinear fit in the appropriate bias range. However, as shown in fig. 4.27a and fig. 4.28b, both methods are in good agreement with the model data only for fairly high bias, where the assumptions made for simplifications hold. Since it is impossible to determine the valid extraction range over bias without prior knowledge of the base resistance, only a single R_B value for each method was determined at the highest measurable bias point and geometry scaling applied.

Similarly, for the calculation of the relative error of all methods, the compact model value $R_\mathrm{B} = R_{\mathrm{Bi}} + R_{\mathrm{Bx}}$ at the highest measurable bias point was used as a reference value.

Two cases are considered for the determination of the relative error: (i) the full model including self-heating and parasitic capacitances, and (ii) an isothermal simulation without the parasitics. Compared to measured data, these cases represent on the one hand a standard measurement and on the other hand an ideal case in which the parasitic capacitances are deembedded based on, e.g., a solution of Poisson's equation with precisely known device geometry, and a pulsed measurement yields (nearly) isothermal data. The relative error for each method is summarized in tab. 4.4 and tab. 4.5. 'CirImI' correspond to the improved method according to (4.3.5-42). Note, that the bias range for simulations including self-heating was limited to the measurable range for real devices in all cases.

According to Table 4.4, the ideal I_B method yields large errors with a random spread. Although the open collector and Z-parameter methods show significantly better results, their errors can still increase to more than 70%. Additionally, both methods give systematically too low values. The standard circle impedance method yields in some cases errors below 20%, but in others still close to 70%, while the improved version reduces the maximum error to around 50%. In order to provide a feel for the geometry dependence of the extracted results and to demonstrate the possible separation of internal and external resistance from geometry scaling, Fig. 4.29 shows for all methods the extracted R_B vs. the emitter mesa width of the GCS test transistors, which were the only width-scaled III-V HBTs available at the time of this writing.

Table 4.5 shows that for ideal measurement conditions the error decreases in most cases. However, even under these ideal circumstances, most methods can still show a considerable relative error. The ideal I_B method remains totally unreliable, and only marginal improvements are observed for the open collector method. The results of the isothermal Z-parameter method agree quite well now (for operation in forward active mode, cf. discussion in sec-

Table 4.4: Relative error (in %) of the base resistance for various extraction methods and different HBT sizes, full model simulation.

Tech	GCS, $l_{E0} = 15\,\mu m$					$b_{E0} = 0.8\,\mu m$			TD	IAF
$A_{E0}\,/\,\mu m^2$	0.5x15	0.8x15	1x15	1.5x15	2x15	0.8x3	0.8x5	0.8x10	0.5x10	0.7x4
ideal I_B	-640	157	-140	-180	-137	-156	-306	-204	-667	477
OpenC	-14	-39	-47	-61	-69	-52	-51	-71	-74	-67
ZPara	-37	-42	-36	-24	-24	-64	-55	-47	-14	-83
HPara	-9	-13	-15	-24	-32	-35	-32	-31	-39	-40
CirIm	-22	-12	-12	-17	-22	-55	-50	-48	-69	-57
CirImI	10	3	3	2	1	-36	-33	-33	-54	-47

Table 4.5: Relative error (in %) of the base resistance for various extraction methods and different HBT sizes. Isothermal simulation and parasitic capacitances omitted.

Tech	GCS, $l_{E0} = 15\,\mu m$					$b_{E0} = 0.8\,\mu m$			TD	IAF
$A_{E0}\,/\,\mu m^2$	0.5x15	0.8x15	1x15	1.5x15	2x15	0.8x3	0.8x5	0.8x10	0.5x10	0.7x4
ideal I_B	-21	61	88	163	222	-348	-128	-148	-2351	648
OpenC	-12	-33	-42	-56	-65	-87	-83	-70	-50	51
ZPara	-11	-9	-8	-7	-11	-3	-3	-3	-5	-6
HPara	-12	-13	-13	-13	-16	-16	-14	-12	14	1
CirIm	-10	-9	-9	-9	-7	-27	-9	-7	-41	-42
CirImI	5	4	5	6	9	-8	13	12	-7	-30

tion II). The circle impedance method also improves in almost all cases under ideal conditions. These results simply show that especially self-heating is the cause for the deterioration of the results under realistic measurement conditions. The extended range permitted by isothermal measurements improves the method accuracy due to better agreement of conditions in the device with the idealized assumptions of the methods.

In fig. 4.29, most methods show a reasonable dependence on geometry. R_{Bx} can be obtained from the y-axis intercept by extrapolation of the straight line, assuming R_{Bi} scales linearly with b_{E0}. The latter dependence is valid for $b_{E0} \ll l_{E0}$ and was implemented in the compact model for the purposes of this comparison, but will differ if b_{E0} approaches l_{E0} [SC10, Sch91a, SKLC08, YZ13]. The relative error of the extrapolation ranges from 35% for the Z-Parameter method to 90% for the uncorrected circle impedance method under

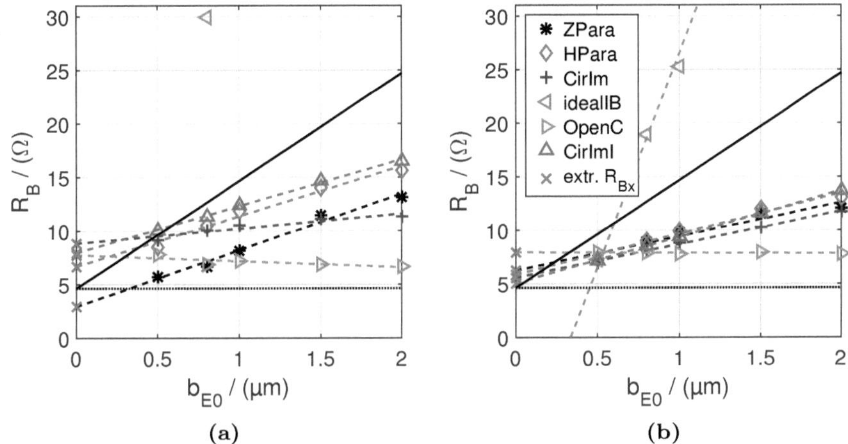

Figure 4.29: Total extracted base resistance vs. emitter mesa width (GCS transistors only) for determining R_{Bx}: (a) Realistic measurement conditions including self-heating and (b) ideal conditions. The model card values for R_{B} (solid line) and R_{Bx} (dotted line) are also shown. The dashed lines represent linear regression fits to the data for obtaining R_{Bx}. The legend in (b) is valid for both figures.

realistic conditions. For isothermal simulations, the relative error improves to 10% for the improved circle impedance method in the best case.

Fig. 4.30 shows results of various extraction methods applied to measured data. Values obtained from the test structure approach have also been included as reference where available (cf. dashed line). While it is impossible to determine the extraction method accuracy based on these data points (due to the lack of a reference), the trend over geometry allows at least a basic statement about the viability and reliability of a method. While a variation of data points can be caused by process tolerances, results that do not or incorrectly correlate with emitter size are likely wrong.

As expected from the comparison to simulated data, the different single transistor based methods show a large, partially random spread including a non-monotonous size dependence. Moreover, they generally underestimate the base resistance determined from the test structures due to partial shorting of R_{Bi}. These results verify the earlier conclusion that base resistance extraction methods are unreliable when applied to the characteristics of single transistors only.

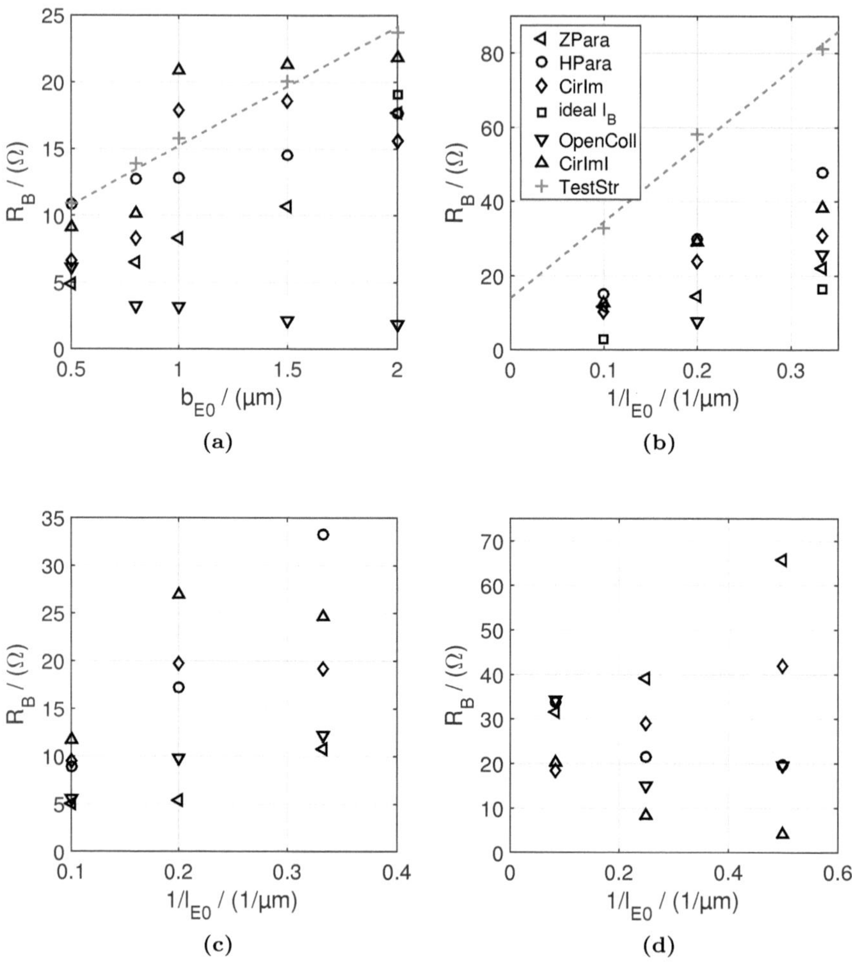

Figure 4.30: Results of extraction methods applied to measured data for the investigated technologies. (a) GCS devices with varying emitter width at constant $l_{\mathrm{E0}} = 15\,\mu\mathrm{m}$; (b) GCS devices with varying emitter length at constant $b_{\mathrm{E0}} = 0.8\,\mu\mathrm{m}$; (c) TD devices and (d) IAF devices, both with different emitter sizes. The legend in (b) applies to all panels.

4.4 Single device parameter extraction

Since at the time of this writing the test structures used for a full HICUM/L2 extraction are not usually available in III-V processes, this section presents an application of the extraction flow in fig. 4.1 to a single device as an example of the most common expected case. The HICUM model equations can be found in, e.g., [SC10], and are not repeated here. A set of model parameters can be extracted based on measurements of a single HBT, though the separation of physical effects is often difficult and the resulting model is likely not at all scalable and valid only for the considered device.

In order to demonstrate the parameter extraction for multiple material systems, a GaAs HBT with $(f_\mathrm{T}, f_\mathrm{max}) = (60,80)$ GHz and $A_\mathrm{E0} = 2.2x22$ µm^2 is chosen here as an example, while the scalable parameter extraction in section 4.5 is based on an InP process. For the purposes of this extraction, it is assumed that the top view of the device, i.e. the area and perimeter of the BC and the BE mesa, are known as $A_\mathrm{E0} = 2.2x22$ µm^2 resp. $A_\mathrm{C0} = 2.8x23$ µm^2, while all other device properties are unknown. The available measured data are based on the measurements listed in tab. 4.1 at four different ambient temperatures. Since the measurements were not obtained over the course of this work, this example extraction is limited to the available data. Where appropriate and available, parameter values are estimated from literature about comparable processes in order to improve the physical basis of the model.

Generally, HICUM/L0 is more suitable for single-transistor extraction due to the reduced parameter set and more simple physical equations (e.g. [NLSD11]). However, given the large distributed BC diode typical for III-V devices and the focus of this work on HICUM/L2, the latter is used in this section and some parameters (relating to high-frequency effects) are disregarded. This section therefore demonstrates that, even with minimal information about a given process, a reasonable parameter set for HICUM/L2 can be obtained.

Following the flow chart in fig. 4.1, the first steps for the extraction are to determine the series resistances, junction diode parameters and the junction and parasitic capacitances.

4.4.1 Junction capacitances

For the junction capacitances, it is assumed that the diode currents and the series resistances are negligible when measuring cold S-parameters at sufficiently low frequencies and the equivalent circuit therefore reduces to the one shown in fig. 4.31a, where C_{BC} is the sum of the BC parasitic and junction capacitance, C_BE is similarly defined for the BE junction and C_CE corresponds

- for III-V HBTs, as discussed in section 3.3 - to the parasitic CE capacitance. Then, the capacitance values can be determined from the Y-parameters as

$$C_{\mathrm{BC}} = \frac{-\mathrm{Im}(\underline{Y_{12}})}{2\pi f} = \frac{-\mathrm{Im}(\underline{Y_{21}})}{2\pi f}, \tag{4.4.1-43}$$

$$C_{\mathrm{BE}} = \frac{\mathrm{Im}(\underline{Y_{11} + Y_{12}})}{2\pi f} \quad \text{and} \tag{4.4.1-44}$$

$$C_{\mathrm{CE}} = \frac{\mathrm{Im}(\underline{Y_{22} + Y_{12}})}{2\pi f}. \tag{4.4.1-45}$$

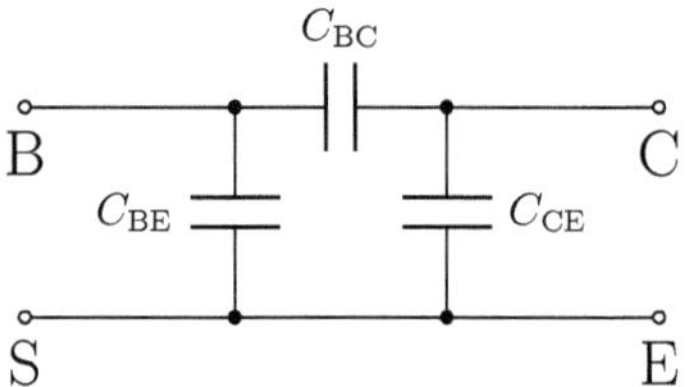

(a)

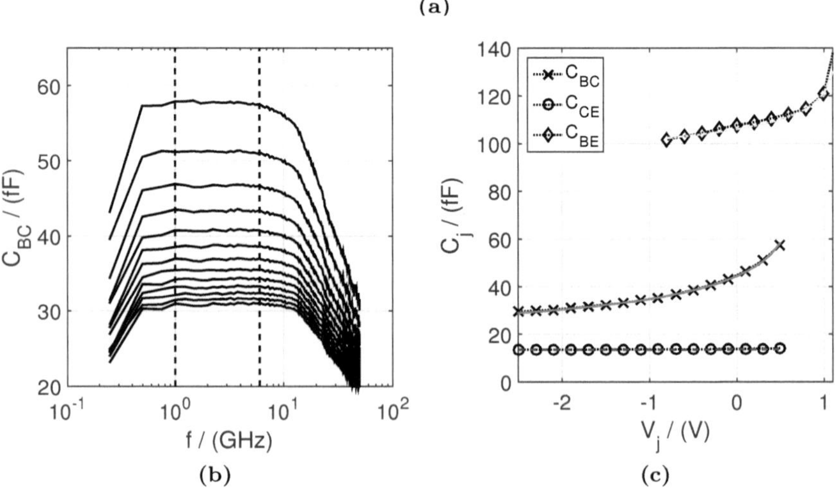

(b)   (c)

Figure 4.31: (a) Equivalent circuit for extracting the junction and parasitic capacitances. (b) Extraction region choice for BC junction capacitance for multiple V_{BC}. Dashed lines indicate recommend extraction region. (c) Extracted (symbols) and modeled (lines) capacitances w.r.t. junction voltages (V_{BC} for C_{jC}, V_{BE} for C_{jE} and $V_{\mathrm{CS}} = V_{\mathrm{CE}}$ for C_{jS}).

The extraction is performed for a single frequency, the spot frequency f_0. f_0 needs to be sufficiently large to obtain a clean measurement signal (i.e. not affected by noise), but sufficiently low to avoid the influence of RC time constants. The calculated capacitance in the extraction region should be constant. An example is given in fig. 4.31b. Here, $f_0 = 3\,\mathrm{GHz}$ was chosen.

For single-transistor extraction, the capacitance parameters are obtained from nonlinear fitting w.r.t. the controlling voltage. The parasitic BE- resp. BC capacitance is considered a fitting parameter. The result for all capacitances is shown in fig. 4.31c. Good agreement is obtained in all cases. Note, that the assumption of a bias-independent C_{CE} capacitance is validated by the measurement.

No extraction vs. the device geometry is performed. However, since the BC diode in III-V mesa HBTs is large and spread out, the BC junction diode zero-bias capacitance is distributed between the internal and the external node according to

$$C_{\mathrm{jCi0}} = \frac{A_{\mathrm{E0}}}{A_{\mathrm{BC0}}} C_{\mathrm{j0,c}} \quad \text{and} \quad C_{\mathrm{jCx0}} = \frac{A_{\mathrm{BC0}} - A_{\mathrm{E0}}}{A_{\mathrm{BC0}}} C_{\mathrm{j0,c}} \qquad (4.4.1\text{-}46)$$

where A_{BC0} is the BC mesa area and $C_{\mathrm{j0,c}}$ is the zero-bias capacitance obtained from the nonlinear curvefit. The remaining model parameters describing the capacitance are equal for the internal and external part, i.e. $z_{\mathrm{Ci}} = z_{\mathrm{Cx}}$ etc. The emitter perimeter component is set to zero.

4.4.2 Diode parameters

The junction diode parameters are obtained from low-bias measurements via linear regression on the rearranged equation for $mV_{\mathrm{T}} >> 1$, i.e.

$$\ln(I_{\mathrm{D}}) = \ln(I_{\mathrm{S}}) + \frac{V_{\mathrm{j}}}{mV_{\mathrm{T}}}, \qquad (4.4.2\text{-}47)$$

with the measured current I_{D} and the model parameters I_{S} and m from the ideal diode equation (e.g. (4.3.3-18)). Where required, the operating range is split into the recombination and the injection range, each with their own parameter set, and the data range for the linear regression is limited to the regions in which one current is dominant. Note, that the upper limit of the extraction range is given by the effect of resistances and self-heating, which cannot be included at this stage.

Fig. 4.32a shows the extraction example for the BE current and fig. 4.32b for the BC current. The BE current appears well-behaved, showing no non-ideal behavior as discussed in, e.g., section 4.3.5.1. The BC current, however, shows some leakage for low voltages. The current is very small and the region is disregarded.

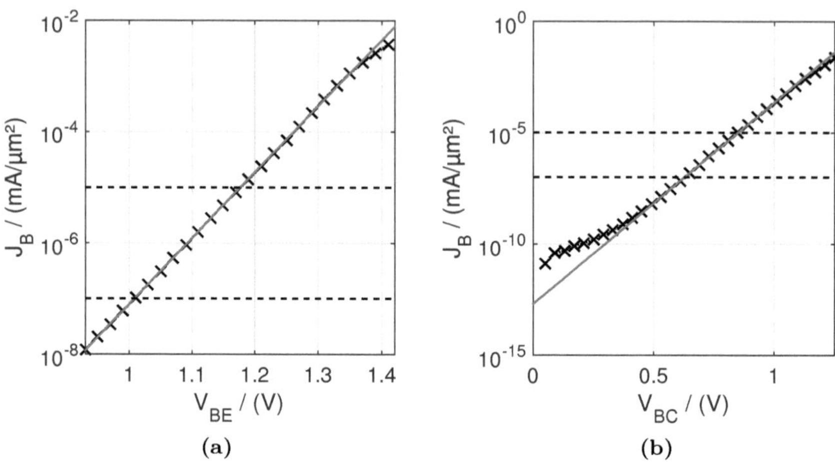

Figure 4.32: Extraction of the model parameters for (a) the BE diode current density at $V_{\mathrm{BC}} = 0\,\mathrm{V}$ and (b) the BC diode current density at $V_{\mathrm{BE}} = 0\,\mathrm{V}$; both at $T = T_0 = 300\,\mathrm{K}$. Symbols: measured data. Lines: diode current according to (4.4.2-47). Dashed lines indicate the extraction region limits.

As in section 4.4.1 for the junction capacitances, the BE perimeter current is set to zero while the BC diode current is split between the internal and external contributions according to the respective area.

4.4.3 Series resistances

Methods for determining the series resistances from single-device measurements were already discussed in section 4.3 and are not repeated here. Given the available measured data and the results of the method accuracy investigation, the Z-parameter method was chosen here for all series resistances. For the emitter resistance, the simultaneous R_{E}-R_{th} method was also applied. The results are given in tab. 4.6. Since the Z-parameter method is known to overestimate the emitter resistance, but the simultaneous method shows deviations in the practical application when the base current is not ideal, the average of both methods was used as an initial value for R_{E}.

The internal and external base resistance cannot be extracted precisely due to the lack of suitable information for the separation. However, from a contact resistance of $R_{\mathrm{c,base}} = 100\,\Omega\,\mu\mathrm{m}^2$ for comparable technologies [Yu70, LDB$^+$94] and a base contact area of approximately $A_{\mathrm{B,con}} = l_{\mathrm{E0}} \cdot 0.6\,\mu\mathrm{m}$ from the size of the BC mesa, a total contact resistance of $r_{\text{con,base}} \approx 7.6\,\Omega$ can be estimated. Then, the remaining extracted resistance is split evenly between R_{Bi} and R_{Bx} due to lack of other information.

Table 4.6: Series resistances obtained by the extraction methods.

Method	R_{Bi}/Ω	R_{Bx}/Ω	R_{Cx}/Ω	R_{E}/Ω
Z-param.	4.9	12.5	3.2	2.91
Simult.	-	-	-	1.8

The parameters for the bias dependence of R_{Bi} are set so that the bias dependence is low, i.e. $f_{\text{dqr0}} = 4$, $f_{\text{qi}} = 1$ and $f_{\text{geo}} = 0$. The zero-bias hole charge q_{p0} is usually extracted from a tetrode test structure and also needs to be approximated here. It is taken from literature information about comparable processes (e.g. [YZK15]) as

$$q_{\text{p0}} \approx q \cdot N_{\text{B}} \cdot w_{\text{B}} \cdot A_{\text{E0}} \approx 28\,\text{pC}. \qquad (4.4.3\text{-}48)$$

4.4.4 Avalanche current parameters

The avalanche current is defined as $I_{\text{AVL}} = -(I_{\text{B}} - I_{\text{BE}})$ at low V_{BE} with the measured base current I_{B} and I_{BE} as the base current extrapolated w.r.t. V_{CE} from the linear range before the onset of self-heating or avalanche, allowing the determination of the multiplication factor $M_{\text{AVL}} = \frac{I_{\text{AVL}}}{I_{\text{C}} - I_{\text{AVL}}}$.

The model describes M_{AVL} as a function of C_{jCi}, since the electric field in the collector can best be described by means of the capacitance [SC10]; the model parameters f_{avl} and q_{avl} are determined from a linear interpolation of the rearranged model equation

$$\ln\left(\frac{M_{\text{AVL}}}{V_{\text{DCi}} - V_{\text{B'C'}}}\right) = \ln(f_{\text{avl}}) - \frac{q_{\text{avl}}}{C_{\text{jCi}}(V_{\text{DCi}} - V_{\text{B'C'}})} \qquad (4.4.4\text{-}49)$$

as the slope and y-intercept, respectively.

However, when attempting to determine the avalanche current from the measured range for the GaAs device, it is observed that it is approximately zero, as shown in fig. 4.33a. Note, that the same conclusion can be drawn for InP HBTs (cf. fig. 4.33b). Therefore, the extraction cannot be demonstrated here and the model parameters are set to zero. In general, the avalanche current is of nearly no relevance for III-V HBTs in the measured V_{CE} range.

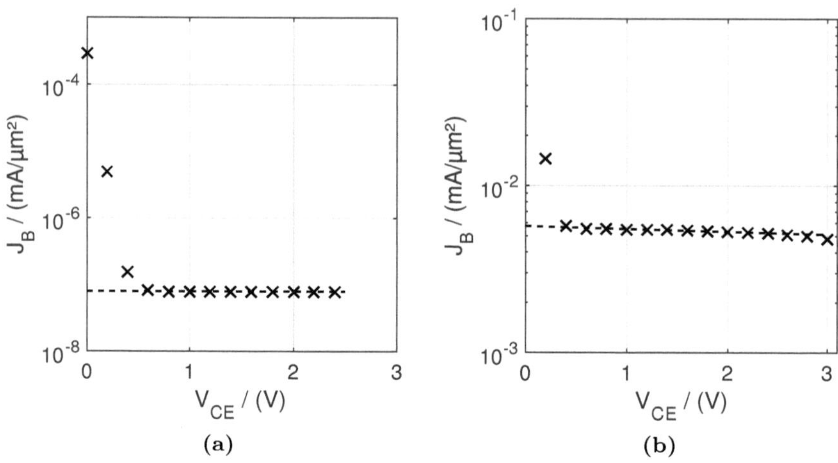

Figure 4.33: Determination of the avalanche current from linear extrapolation (lines) of measured J_{BE} (symbols) w.r.t V_{CE} for (a) a GaAs HBT and (b) an InP HBT. The former does not show a deviation of the base current from the linear extrapolation, the latter shows only the impact of the leakage current as discussed in section 5.2, demonstrating that the avalanche effect can be neglected for modeling in the measured voltage range.

4.4.5 Temperature coefficients and thermal resistance

The diode current temperature dependence parameters are determined by applying the extraction as described in section 4.4.2 at multiple ambient temperatures, keeping in mind that V_T is temperature dependent and forcing the nonideality factor m to the same value as for T_0. Then, the model parameters are obtained by nonlinear fitting on the temperature dependence of the saturation current

$$I_{xS}(T) = I_{xS}(T_0) \left(\frac{T}{T_0} \right)^{\zeta_x} \exp\left[\frac{V_{gx}}{V_T} \left(\frac{T}{T_0} - 1 \right) \right]. \tag{4.4.5-50}$$

In this way, the parameters V_{gE}, V_{gC}, ζ_{BET}, ζ_{Ci} and ζ_{Cx} are determined. In the single transistor extraction, $\zeta_{Ci} = \zeta_{Cx}$ since the contributions cannot be separated. The data for the GaAs process is shown in fig. 4.34a.

The parameters f_{1vg}, the temperature dependence of the band gap, and V_{gBE} would typically be extracted from the temperature dependence of the BE junction capacitance. Here, though, the saturation current I_{REs} over

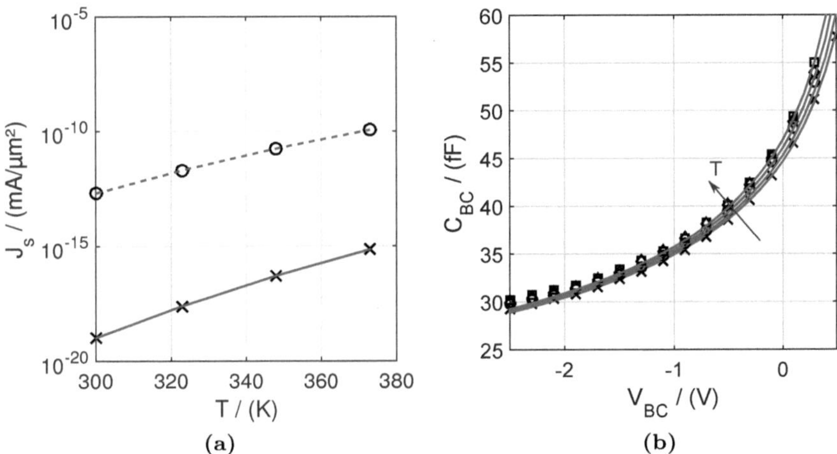

Figure 4.34: Example for the determination of the temperature dependence parameters for (a) the diode saturation currents (BE junction: x, solid line; BC junction: o, dashed line) and (b) the BC junction capacitance for $T = [300, 325, 350, 375]$ K. Symbols represent the extracted resp. measured data points, lines the temperature-dependent model.

temperature would be preferred. The reason for this preference is that capacitance data over temperature is often noisy even if measured at relatively high input power, most likely due to the change of the calibration standards over temperature. Another reason is the large deembedding capacitance compared to the junction capacitance [Ros16]. Since, however, the recombination current cannot be found in the measured data, f_{1vg} is extracted along with V_{gBC} based on a nonlinear fit on C_{BC} as in section 4.4.1 for multiple ambient temperatures from

$$C_{jo}(T) = C_{jo}(T_0) \left[\frac{V_D(T_0)}{V_D(T)} \right]^z \quad \text{and} \tag{4.4.5-51}$$

$$V_D(T) = V_D(T_0) - m_g V_T \ln\left(\frac{T}{T_0}\right) - V_g\left(\frac{T}{T_0} - 1\right) \quad \text{with} \tag{4.4.5-52}$$

$$m_g = 3 - \frac{q f_{1vg}}{k_B} \tag{4.4.5-53}$$

while keeping the parameters extracted at T_0 constant.

V_{gBE} is based on the same procedure for C_{BE}, taking f_{1vg} as given. The example for the GaAs HBT is shown in fig. 4.34b.

The temperature dependence parameters of the series resistances ζ_{re}, ζ_{rcx}, ζ_{rbi} and ζ_{rbx} are obtained by applying the extraction methods as in section 4.4.3 for multiple ambient temperatures and fitting the equation

$$R(T) = R(T_0) \cdot \left(\frac{T}{T_0}\right)^{\zeta}. \tag{4.4.5-54}$$

Note, that $\zeta_{rbi} = \zeta_{rbx}$ due to lack of other information. Data for R_{Cx} were too noisy to obtain reliable values for multiple temperatures; ζ_{rcx} was set to zero. Examples are shown in fig. 4.35.

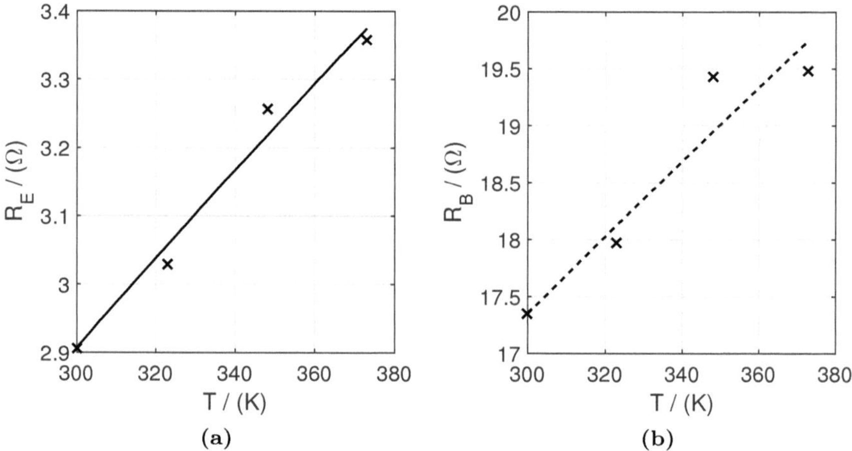

(a)   (b)

Figure 4.35: Example for the determination of the temperature dependence parameters for (a) the emitter resistance and (b) the base resistance capacitance for $T = [300, 325, 350, 375]$ K, both using the Z-Parameter method. Symbols represent the extracted data points, lines the temperature-dependent model.

Having obtained the temperature dependence of the diode currents, the thermal resistance can be extracted. Under the assumption that the entire BE current flows through a single diode with a single set of temperature parameters and given a constant collector current, which can typically be obtained in III-V devices by measuring a V_{CE} sweep with forced I_B, the thermal resistance can be determined from the bias dependence of V_{BE} w.r.t. $P \approx I_C V_{CE} + I_B V_{BE}$ using

$$I_{BEis,T}(T) = I_{BEiS} \left(\frac{T}{T_0}\right)^{\zeta_{BET}} \exp\left(\frac{V_{gE}}{V_T}\left[\frac{T}{T_0} - 1\right]\right), \tag{4.4.5-55}$$

$$V_{\mathrm{B'E'}} = m_{\mathrm{BEi}} V_{\mathrm{T}} \ln\left(\frac{I_{\mathrm{BE}}}{I_{\mathrm{BEis,T}}}\right) + R_{\mathrm{E}} I_{\mathrm{E}} + R_{\mathrm{B}} I_{\mathrm{B}} \qquad (4.4.5\text{-}56)$$

and

$$T = T_0 + P \cdot R_{\mathrm{th}}, \qquad (4.4.5\text{-}57)$$

and therefore

$$\frac{\mathrm{d}V_{\mathrm{B'E'}}}{\mathrm{d}P} \approx \frac{\mathrm{d}V_{\mathrm{B'E'}}}{\mathrm{d}T}\frac{\mathrm{d}T}{\mathrm{d}P} = \frac{m_{\mathrm{BEi}} V_{\mathrm{T0}}}{T_0} R_{\mathrm{th}} \left[\ln\left(\frac{I_{\mathrm{B}}}{I_{\mathrm{BEiS}}}\right) - \frac{V_{\mathrm{gE}}}{V_{\mathrm{T}}}\frac{T_0}{R_{\mathrm{th}}P + T_0} - \zeta_{\mathrm{BET}}\right].$$
$$(4.4.5\text{-}58)$$

This neglects the temperature dependence of the base, emitter and thermal resistances, which is generally permissible when using small values for I_{C}, as well as the small dependence of the dissipated power on the actual BE voltage. Further assuming that the term $\frac{T_0}{R_{\mathrm{th}}P + T_0} \approx 1$ for sufficiently low self-heating permits the direct extraction of R_{th} from linear regression of V_{BE} w.r.t. P. Here, the full equation was used in a least-squares fit.

For comparison and verification, R_{th} was also determined with the two simultaneous R_{E}-R_{th} methods as investigated in section 4.3.3. Both results were within 10% of the value found using (4.4.5-58). Note, that both extraction methods make similar assumptions as the extraction shown here, e.g. neither method permits a distributed current flow.

An example for the extraction is shown in fig. 4.36a, with fig. 4.36b showing the verification for the assumed constant collector current as well as the definition of the extraction range.

The thermal resistance is itself a temperature-dependent parameter. It is extracted at multiple ambient temperatures and described as

$$R_{\mathrm{th}}(T) = R_{\mathrm{th}}(T_0) \cdot (1 + \alpha_{\mathrm{Rth}}(T - T_0)). \qquad (4.4.5\text{-}59)$$

The thermal capacitance C_{th} can be extracted from low-frequency measurements if the impact of the input power of the S-parameter signal is visible in the measured data. Since no low-frequency measurements were performed, C_{th} is set for an assumed thermal RC time constant of $\tau_{\mathrm{th}} = 300\,\mathrm{ns}$ (cf., e.g., section 4.2.3) as

$$C_{\mathrm{th}} \approx \frac{\tau_{\mathrm{th}}}{R_{\mathrm{th}}} \approx 0.28\,\mathrm{nF}. \qquad (4.4.5\text{-}60)$$

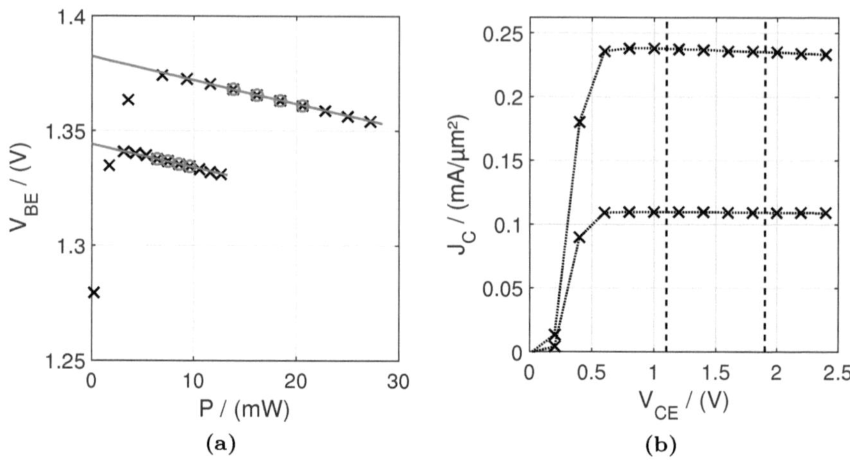

Figure 4.36: (a) Extraction of R_{TH} from linear regression of V_{BE} w.r.t. P. Crosses indicate the measured voltage, circles the extraction region chosen according to constant I_{C}. (b) Collector current w.r.t V_{CE} for forced I_{B}. The dashed lines indicate the extraction region choice.

4.4.6 Transit time parameters

The transit time is classically determined from

$$\tau_{\mathrm{f}} = \frac{1}{2\pi f_{\mathrm{T}}} - \left((R_{\mathrm{E}} + R_{\mathrm{Cx}}C)_{\mathrm{BC}} - \frac{C_{\mathrm{BB}}}{g_{\mathrm{m}}} \right), \tag{4.4.6-61}$$

with the sum of all capacitances at the transistor base node C_{BB} and the transconductance g_{m}. The transit frequency f_{T} is determined as

$$f_{\mathrm{T}} = \frac{f_{\mathrm{meas}}}{\mathrm{Im}\left(\frac{Y_{11}}{Y_{21}} \right)}, \tag{4.4.6-62}$$

and similar considerations for the valid frequency range as for the junction capacitance determination in section 4.4.1 apply.

A more precise approach, however, is to use the parameters determined to deembed the external and parasitic elements using two-port matrices [RSPL13]. The equivalent circuit of the transistor in emitter configuration and the definition of the two-port matrices are given in fig. 4.37, neglecting the substrate transistor and diode, which are not active during the extraction of the τ_f parameters. Note, that the internal base resistance cannot be deembedded in the same fashion due to the transconductances resulting from its voltage dependence. Fortunately, it has little to no impact on $\underline{Y_{11}}$ and $\underline{Y_{21}}$ of the internal transistorand thus on the determination of f_T in the measured frequency range.

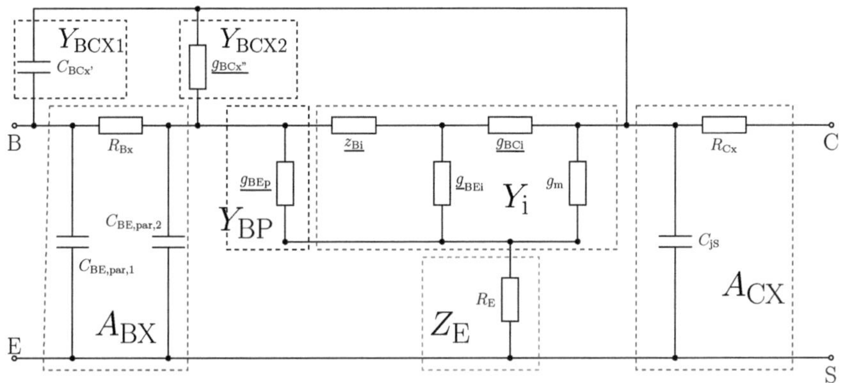

Figure 4.37: HICUM small-signal equivalent circuit neglecting substrate elements and sub-circuits showing the two-port matrices used for deembedding the external, peripheral and parasitic elements from the internal transistor. For clarity, complex conductances were used for the junctions where appropriate.

The deembedding is performed as

$$Y_\mathrm{i} = z2y(y2z(a2y(A_\mathrm{BX}^{-1} \cdot y2a(a2y(y2a(Y_\mathrm{meas}) \cdot A_\mathrm{CX}^{-1}) - Y_\mathrm{BCX1})) - Y_\mathrm{BCX2}) - Z_\mathrm{E}) - Y_\mathrm{BP}, \tag{4.4.6-63}$$

with, e.g. $z2y$, as the transformation of a Z-matrix to a Y-matrix and the other transformations similarly defined. (4.4.6-63) allows a precise determination of the internal Y-parameters. Note that $Y_\mathrm{BP} = 0$ for this extraction.

All extraction steps for the remainder of this section are based on the internal Y-parameters. Similarly, for voltage-dependent quantities like the critical current, the internal device voltages are used based on the measured currents and the extracted resistances.

The low-bias transit time is determined from extrapolation of $1/(2\pi f_\mathrm{ti})$ vs. $1/I_\mathrm{C}$, as shown in fig. 4.38a. The parameters t_0, d_t0h and t_bvl are determined from the nonlinear fit of

$$\tau_\mathrm{f0} = t_0 + d_\mathrm{t0h}(c - 1) + t_\mathrm{bvl}(\frac{1}{c} - 1), \tag{4.4.6-64}$$

with the inverse internal BC junction capacitance normalized to its zero-bias value $c = \dfrac{C_\mathrm{jCi0}}{C_\mathrm{jCi}}$. Simultaneously, the current-dependent collector transit time $\Delta\tau_{\mathrm{C}}$ is determined from the *decrease* of the transit time vs. the linear extrapolation, which does not occur in SiGe HBTs (cf. section 3.5). The current-independent collector transit time cannot be separated from τ_{f0} and is included in the fit. The equations for the collector transit time are as given and discussed in section 3.5. The low-current transit times as determined from measurements and model are shown in fig. 4.38b.

At the time of this writing, only the semi-empirical model as described in section 3.5.2.2 for the current dependence of $\Delta\tau_{\mathrm{C}}$ was available, using a smooth transition to zero of the transit time with increasing collector current and a linearly voltage-dependent transition current I_{NDM}. The transition current for a given voltage is determined from nonlinear fitting of the collector transit time normalized to its zero-current value, as demonstrated in fig. 4.38c. The extraction of the parameters I_{NDM0} and k_{NDM} from linear regression on the determined current is shown in fig. 4.38d. Note, that $\tau_{\mathrm{c}}(J_{\mathrm{C}})$ must be subtracted from the linear extrapolation from τ_{f} w.r.t. $1/J_{\mathrm{C}}$ for the determination of $\Delta\tau_{\mathrm{f}}$, since the latter models the increase of the transit time w.r.t. its minimum, i.e. $\Delta\tau_{\mathrm{f}}$ is not the difference between the linear extrapolation and the transit time as shown in fig. 4.38a, but rather the difference to the same extrapolation corrected by $\Delta\tau_{\mathrm{C}}$ and therefore passing through the transit time minimum. $\Delta\tau_{\mathrm{f}} >= 0$ holds for all bias conditions.

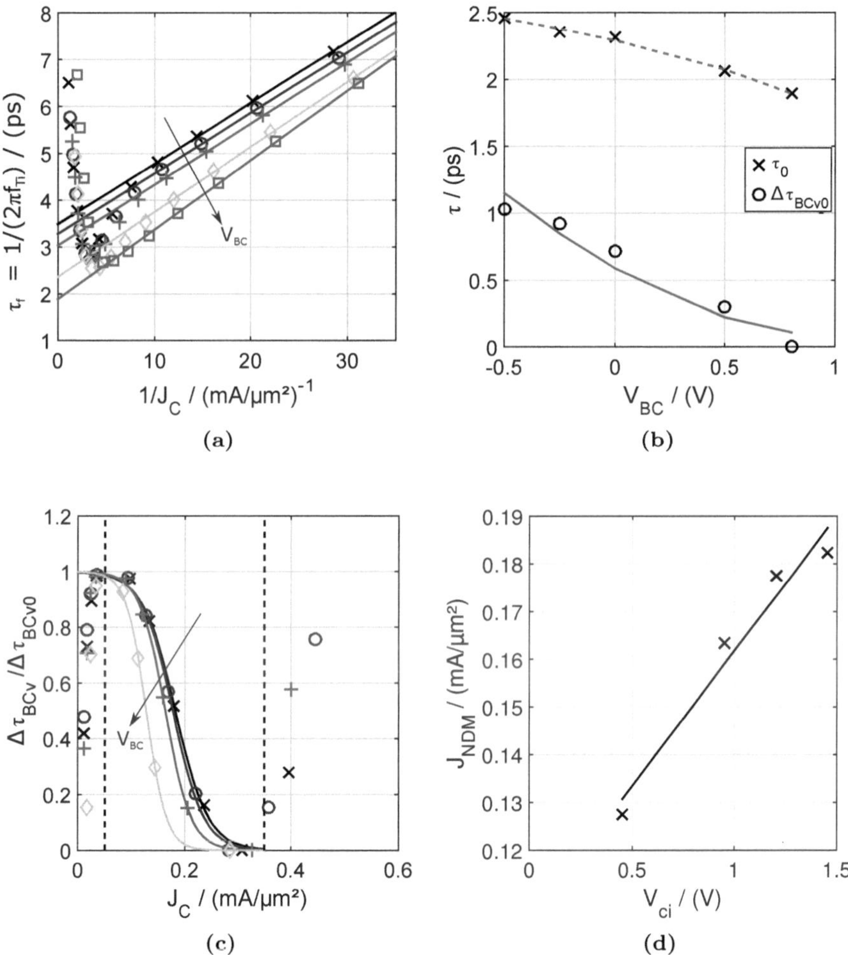

Figure 4.38: (a) Linear extrapolation of τ_{f0} and $\Delta\tau_C$ from internal τ_f vs. $1/J_C$ for $V_{BC} = [-0.5, -0.25, 0, 0.5, 0.8]$ V; (b) resulting values (symbols) and model fit (lines) for both; (c) current dependence of the collector transit time from measurements (symbols) and model (lines); dashed lines indicate the extraction range. (d) Extracted J_{NDM} vs. internal collector voltage (symbols) and model interpolation (line).

Having determined $\Delta\tau_\mathrm{f}$, the critical current for the given GaAs HBT is defined as $I_\mathrm{CK} = I_\mathrm{C}(\Delta\tau_\mathrm{f} = 1\,\mathrm{ps})$ and extracted for different V_CE or V_BC. Then, the parameters for the critical current r_Ci0, v_lim, v_ces, v_pt, a_ick and del_ck can be obtained from nonlinear fitting of

$$I_\mathrm{CK}(V_\mathrm{C'E'}) = \frac{v_\mathrm{c,eff}}{r_\mathrm{ci0}} \cdot \frac{1 + \frac{1}{2}\left(v + \sqrt{v^2 + a_\mathrm{ick}}\right)}{\left[1 + (v_\mathrm{c,eff}/V_\mathrm{lim})^{del_\mathrm{ck}}\right]^{1/del_\mathrm{ck}}}, \tag{4.4.6-65}$$

on the measured data with

$$v_\mathrm{c,eff} = V_\mathrm{T}\left[1 + \frac{u + \sqrt{u^2 + 1.921812}}{2}\right] \tag{4.4.6-66}$$

and the shorthand notations

$$u = \frac{V_\mathrm{C'E'} - v_\mathrm{ces} - V_\mathrm{T}}{V_\mathrm{T}} \quad \text{and} \quad v = \frac{v_\mathrm{c,eff} - v_\mathrm{lim}}{v_\mathrm{PT}}. \tag{4.4.6-67}$$

Note, that while a_ick is accessible as a parameter, it was left at its standard value of $a_\mathrm{ick} = 10^{-3}$ here. The critical current density as determined from measurements and model is shown in fig. 4.39a.

Finally, the parameters for the medium- and high-current transit time must be found. It is bias-dependent only via the ratio of $i = I_\mathrm{C}/I_\mathrm{ck}$ and should, if all previous extraction steps are correct, be voltage-independent when plotted against this ratio. This is reasonably accurate for the GaAs HBT extraction, as shown in fig. 4.39b. Then, the components of the medium current emitter transit time $\Delta\tau_\mathrm{Ef}$ and the high-current transit time τ_hcs can be determined from nonlinear fitting.

For this extraction, however, it was sufficient to use only $\Delta\tau_\mathrm{Ef}$, since the measurement data does not extend towards sufficiently high I_C for τ_hcs to become relevant. It appears likely that this remains true until thermal breakdown. The result of nonlinear fitting on the equation

$$\Delta\tau_\mathrm{Ef} = \tau_\mathrm{ef0}\left(\frac{I_\mathrm{C}}{I_\mathrm{ck}}\right)^{g_\mathrm{tfe}} \tag{4.4.6-68}$$

is shown in fig. 4.39b. τ_hcs is set to zero.

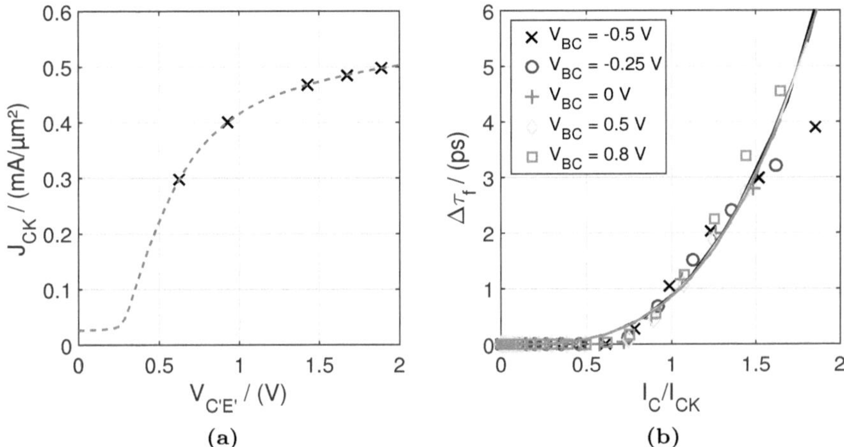

Figure 4.39: (a) Critical current density and (b) high-current transit time $\Delta\tau_\mathrm{f}$ determined from measurements (symbols) and model (lines). In (b), the data for multiple V_BC are shown to be reasonably independent of V_BC.

4.4.7 Transfer current parameters

Like the transit time parameter determination, the extraction of the transfer current parameters consists of a series of interdependent steps for each of the weight factors and the bias dependence of h_jEi.

For SiGe HBTs, the standard extraction sequence, given here for completeness, is as follows [Paw14]:

- Extract the bias dependence parameter a_hjei of the reverse early effect from four bias points with V_BE and ΔV_BE sufficiently large to avoid measurement noise, but sufficiently small to avoid the impact of self-heating and high-current effects.

- Extract the weight factor for the reverse early effect h_jei0 and the GICCR weight factor c_{10} from a linear regression on the inverse normalized transfer current $I_\mathrm{C,n}^{-1} = \left(\dfrac{I_\mathrm{C}}{\exp(V_\mathrm{B'E'}/V_\mathrm{T}) - \exp(V_\mathrm{B'C'}/V_\mathrm{T})} \right)^{-1}$ w.r.t. $h \cdot Q_\mathrm{jEi}$ for $V_\mathrm{BC} = 0$.

- The temperature dependence of the above parameters, ζ_CT, ζ_vgbe, ζ_hjei, V_gB and ΔV_gBE, is determined by applying the steps at different ambient temperatures.

- The weight factor for the early effect h_{jCi} is determined from the slope of the normalized transfer current w.r.t. Q_{jCi} for constant V_{BE}.

- Extraction of the mobile charge weight factor h_{f0} by calculating the medium-bias weighted hole charge as $Q_{\mathrm{pT}} = \frac{c_{10}\exp(V_{\mathrm{B'E'}}/V_{\mathrm{T}})}{I_{\mathrm{C}}}$, removing the known weighted junction charges and the zero-bias hole charge and linearly extrapolating towards $I_{\mathrm{C}} = 0$. Note, that self-heating needs to be considered for all components.

- The weight factor h_{fE} is extracted from the weighted hole charge Q_{pT} at high bias by removing the weighted mobile charge and calculating the ratio of the remaining term to the emitter charge ΔQ_{Ef}.

The extraction of the of the weight factor for the reverse early effect is based on solving the implicit equation

$$\frac{h_1 Q_{\mathrm{jEi},1} - h_2 Q_{\mathrm{jEi},2}}{h_3 Q_{\mathrm{jEi},3} - h_4 Q_{\mathrm{jEi},4}} - \frac{\dfrac{\exp(V_{\mathrm{B'E'}1}/V_{\mathrm{T}})}{I_{\mathrm{C},1}} - \dfrac{\exp(V_{\mathrm{B'E'}2}/V_{\mathrm{T}})}{I_{\mathrm{C},2}}}{\dfrac{\exp(V_{\mathrm{B'E'}3}/V_{\mathrm{T}})}{I_{\mathrm{C},3}} - \dfrac{\exp(V_{\mathrm{B'E'}4}/V_{\mathrm{T}})}{I_{\mathrm{C},4}}} = 0 \qquad (4.4.7\text{-}69)$$

with

$$h_n = \frac{h_{\mathrm{jEi}}(V_{\mathrm{B'E'}n})}{h_{\mathrm{jEi0}}} = \frac{\exp\left[a_{\mathrm{hjEi}}\left(1 - (1 - V_{\mathrm{B'E'}n}/V_{\mathrm{DEi}})\right)\right] - 1}{a_{\mathrm{hjEi}}\left(1 - (1 - V_{\mathrm{B'E'}n}/V_{\mathrm{DEi}})\right)} \qquad (4.4.7\text{-}70)$$

for a_{hjEi} [Paw14].

For III-V HBTs, the standard extraction procedure for h_{jEi} usually fails to generate a satisfactory set of parameters. Two major reasons for this are (i) the difficulty in determining the emitter resistance with sufficient precision and the subsequent impact on the calculated internal BE voltage and (ii) the presence of small leakage currents. The latter are usually very small and do not affect the operating regions relevant for circuit design, but may be large enough to affect, e.g., the slope of the normalized collector current in the low-current region required for the extraction. Besides those difficulties, though, from experience based on the parameter extraction for multiple III-V processes, the weight factors related to the mobile charge h_{fe}, h_{fc} and h_{f0} as well as the early effect h_{jci} are nearly irrelevant for III-V HBTs; the BE heterojunction dominates the transfer current. Therefore, it makes sense that their exact value is extremely difficult to extract. All of those parameters can be set to fixed values $\ll 1$.

Then, the initial value for a_{hjEi} can be determined as described above. With a given or estimated Q_{p0}, the parameter c_{10} can be extracted from a linear extrapolation of $\ln(I_{\mathrm{C}})$ w.r.t V_{BE} in the low-bias region, in analogy to the diode currents in section 4.4.2, with the temperature dependence being similarly determined from the measurements at various ambient temperatures. The weight factor h_{jEi}, the final value for a_{hjEi} as well as ζ_{vgbe}, ζ_{hjei} and ΔV_{gBE} are extracted from least-square fitting of a self-consistent solution for the transfer current in the medium current region first for T_0 then for various T_{amb}. The result of this fit is shown in fig. 4.40.

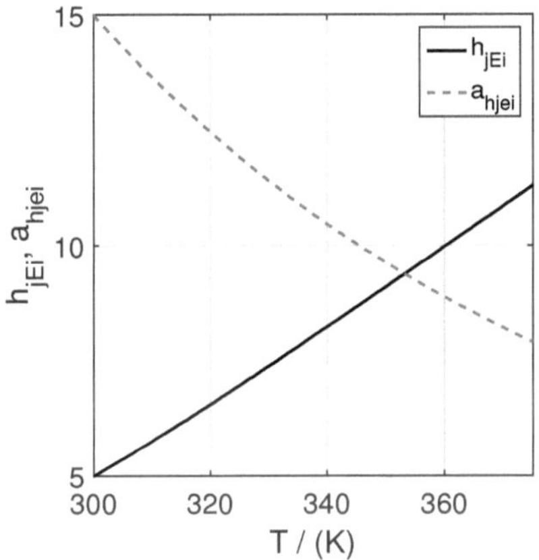

Figure 4.40: Temperature dependence of h_{jEi} and a_{hjei} as determined from concurrent self-consistent fitting.

4.4.8 Additional extraction steps

The remaining extraction steps are related to the substrate transistor, which is not relevant for III-V technologies, and to the non-quasi-static effects, which are relevant near or beyond f_{t}. Neither set of steps is thus shown here; if a given technology includes the effects, the extraction is described in, e.g., [Paw14, Kra15, Ros16].

Note, however, that a significant amount of manual optimization is required for a good agreement between model and measurement; the extraction steps described up to this point offer good starting values, but even small discrepancies between the actual and model series resistances can have a major impact on the transit time extraction or the weight factors of the transfer current. The steps as described are still necessary to generate the starting values for the optimization, since modern compact models are too complicated to extract a parameter set by simple curve fitting.

4.4.9 Summary

The extraction of a parameter set for the HICUM compact model based on a measurement set for a single transistor with limited information about the technology was demonstrated here. The comparison of model and measurements is shown in chapter 5. The steps of the extraction procedure were briefly illustrated and the relevant equations given in order to provide an extraction guide specifically for III-V HBTs. The final results of the extraction, along with those for InP HBTs, are shown in chapter 5 and demonstrate excellent agreement.

It has also been shown that the current-dependent collector transit time can be found in measurements.

Even for these less-than-ideal circumstances, it is possible to determine a reasonably accurate parameter set when taking into account not only the proper extraction routines for each parameter, but also making use of additional resources such as literature and the fact that physics-based parameters can often be estimated well due to their underlying connection to the technology. There is, thus, no reason to not deliver physics-based compact models as part of a PDK.

4.5 Geometry-scalable parameter extraction

Though the principle of geometry-scalable modeling has been described elsewhere (e.g. [SC10, Kra15, LWZ$^+$11]), its application to III-V HBTs is, with few exceptions (e.g. [HZC$^+$07]), very rare. Therefore, this section discusses the parameter extraction based on a set of test structures and transistors with multiple emitter geometries, which allow the determination of the internal and peripheral resp. external component for many of the equivalent circuit elements. The most important test structures and their evaluation is presented in section 4.5.2. Section 4.5.3 demonstrates the regional separation. Typical problems occurring during the extraction are also discussed in section 4.5.3.1. Tab. 4.7 lists a minimum recommended set of structures for a scalable HICUM/L2 extraction. The extraction of model parameters for a reference device, e.g., for the transfer current, was already shown in section 4.4 and will not be discussed again.

While HBT scaling can easily become very complicated, depending on the number of different devices with different contact configurations, finger numbers etc. that are to be modeled and would then require a dedicated scaling tool for the creation of the compact model parameters based on the technology parameters (e.g. [ZMWS09]), in this work, only scaling according to the emitter and collector mesa area and perimeter is considered. More detailed scaling rules are commonly the result of several years of incremental modeling work and are at this point not available for the investigated InP HBTs. An example for a scalable model parameter extraction for a mature HBT process with full access to test structures and process information can be found in, e.g., [Ros16].

Table 4.7: Recommended structure list for scalable HICUM/L2 parameter extraction

No	Structure
1	**Tetrode resp. walled-E structures** Transistors with two independent base contacts for measuring the internal base resistance. At least three different transistor widths, preferably for two different transistor lengths each.
2	**TLM structures** Long, wide transistor layers with contacts for direct measurements for determining the sheet resistance of each relevant layer. At least two different lengths, each for two different widths, for each relevant layer with a separate sheet resistance.
3	**Large-area diode structures** Large-area (min. $A_j = 20 \cdot 20$ µm^2), low-perimeter (i.e. typically square) for measurement of internal area dependent junction capacitances. Open deembedding structure (contact metal without semiconductor layers) for each also required.
4	**Thermal coupling structures** Multi-finger transistors (five fingers minimum) with common base and collector contacts, independent emitter contacts for determining the thermal coupling coefficients. Three total, one at reference length and width, one variation of length and width each.
5	**FIB/SEM structures** Multi-finger transistors with different emitter widths and/or lengths for each finger for determining actual vs. drawn geometries. One at constant emitter length with width variation, one at constant width with length variation.
6	**Standard Transistors** Transistors in GSG pads for DC and RF measurements. At least four different l_{E0} at $b_{E0,ref}$ and four different b_{E0} at $l_{E0,ref}$. Open and short deembedding structures for reference device also required.
7	**Other transistor configurations** Transistors in GSG pads with other transistor configurations (multi-finger, contact location etc.) for verification of scaling rules.
8	**On-wafer calibration** Depending on desired calibration method, typically lines of different length for multiline TRL calibration, for high-frequency measurements.

4.5.1 Technology description

Over the course of this work, two sets of test structures, both for InP HBT processes, one with $(f_T, f_{max}) = (300, 250)$ GHz [GCS] and the other with $(f_T, f_{max}) = (350, 500)$ GHz [GDR$^+$05], were evaluated. Since the latter suffers from design problems that make the evaluation of several structures impossible, examples in this section are shown for the former process.

The transistors are structured in the typical mesa cross-section as shown in fig. 2.3. Available transistor sizes are listed in tab. 4.8. Additionally, the following test structures are available:

- Large-area BE and BC diode structures,

- contact chain structures for the base and collector sheet resistances (often referred to as TLM structures in literature),

- thermal coupling structures,

- walled-E structures with multiple emitter widths,

- structures for the dynamic collector resistance,

- multi-finger transistors.

Since scaling in this work is limited to single-finger transistors and area- and perimeter-dependent scaling, the latter structures were not evaluated. Furthermore, thermal coupling is disregarded in this work, since it is not taken into account at the device level. Finally, the large-area diode structures were not used for the model parameter extraction due to the uncertainties regarding device sizes as discussed in section 4.5.3.1.

The results presented in this section were also published in part in [NSC$^+$13, NKPS16].

Table 4.8: Available transistors in GSG pads for the GCS 250 GHz InP technology. The nominal emitter width is 0.8 µm

$b_{E0}/\mu m$	$l_{E0}/\mu m$			
	3	5	10	15
0.5	x			x
0.8	x	x	x	x
1.0	x	x	x	x
1.5	x			x
2.0	x			x

4.5.2 Test structures

In the context of this work, "test structure" refers to any structure fabricated in addition to a conventional transistor for the purposes of device characterization and model parameter extraction. The term sometimes includes all structures used for the latter purpose, but standard transistors are excluded from the definition here in order to separate the parameter extraction based on only transistor measurements from those in which individual effects can be separated. Of the structures listed in tab. 4.7, no. 1-5 are considered as test structures.

Test structures are typically fabricated and contacted using DC pads and probes. In order to remove the contact resistance from measurements, the force-sense method is used in which two probes are used per pad, one carrying the current, the other measuring the voltage while forcing the current flowing through it to zero.

This section shows evaluation examples for the tetrode, sheet resistance structures and dynamic collector resistance structures; the large-area diodes are not as important for III-V devices as for SiGe due to (a) very uniform transistor layers due to epitaxial layer growth and (b) scaling deviations as discussed in section 4.5.3.1. Thermal coupling is not included in the compact model for single transistors but would become relevant for circuits with multiple devices or the effective model for a multi-finger device; neither is considered in this work. Finally, FIB/SEM structures were proposed as a consequence of the fist scalable InP models and have not been fabricated yet. Determining sheet resistances from test structures and geometry calculation, but extracting other model parameters from transistor measurements is comparable to industrial best-practice examples [ZHY$^+$07].

4.5.2.1 Sheet resistance structures

Structures for determining the sheet resistance of a device layer are also called TLM (<u>t</u>ransfer <u>l</u>ength <u>m</u>ethod) structures. They are available for most processes since they are commonly used as process control monitors (PCMs), though they are not usually included on chips after separation. The voltage drop of a current forced through two of the contacts is measured at the second and third contact, allowing the determination of the base contact resistance. Furthermore, by subtracting the measured resistor R_{12} between contact 1 and 2 from R_{23}, the external sheet resistance follows directly if $b_2 = 2b_1$ and $b_1 = l$. Note, that in an III-V process the contact chain structure for the base resistance cannot be fabricated without a collector mesa, which necessitates the addition of a collector contact to ensure $V_{BC} = 0\,\mathrm{V}$. Fig. 4.41a shows a typical implementation of a sheet resistance structure. Additional lengths

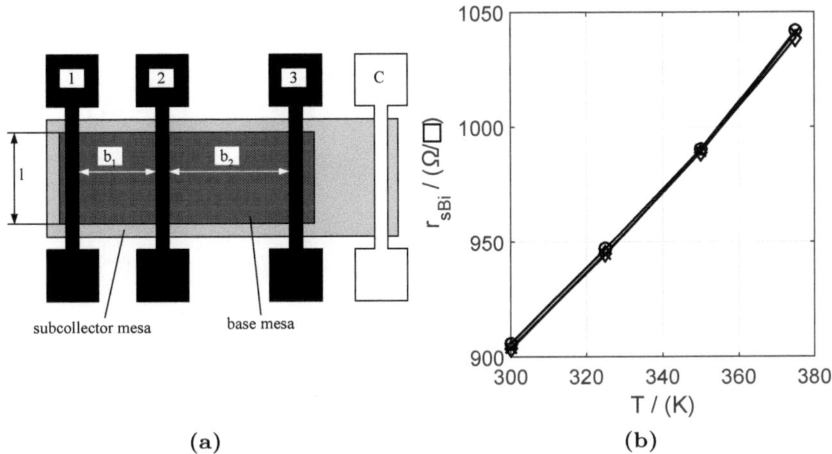

(a)   (b)

Figure 4.41: (a) Schematic view of a TLM structure implemented in a III-V technology, here for the base sheet resistance. (b) Result of TLM structure evaluation for multiple temperatures, here for three structures on the same wafer, showing good repeatability.

can be added as needed to increase reliability.

If structures with different widths and lengths are available, the foreside component of the sheet resistance can also be deembedded by using

$$R_{x,1D} = (G_{2x} - G_{1x}).^{-1} \qquad (4.5.2\text{-}71)$$

with the 1D resistance of the TLM structure $R_{x,1D}$ with the length $\Delta l = l_2 - l_1$. the measured conductances G_{2x} and G_{1x} for l_2 and l_1, respectively, and $x = \{1, 2\}$ for two different b_x. Then, the sheet resistance is

$$R_s = (R_{2,1D} - R_{1,1D}) \cdot \frac{\Delta l}{\Delta b}. \qquad (4.5.2\text{-}72)$$

Using multiple ambient temperatures, the temperature dependence of the resistance can be determined as well. The application example in fig. 4.41b shows good repeatability for multiple structures from the same wafer and for different temperatures.

Note, that the base sheet resistance is not necessarily constant over the entire device; a passivation layer, for example, as is sometimes used to suppress a surface recombination current between the internal base and the base contact, may lead to a different sheet resistance, though at this point, no investigation with regards to this effect are known. For SiGe devices, however,

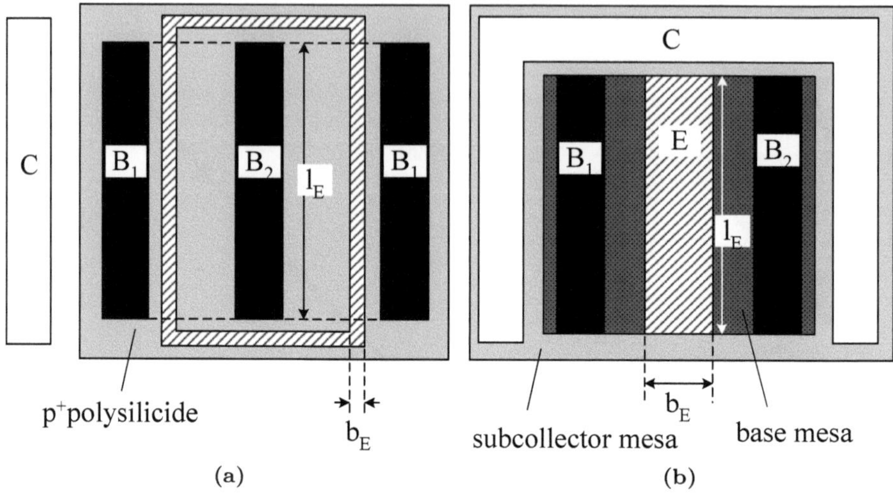

Figure 4.42: Schematic view of (a) a tetrode ring structure as commonly implemented in SiGe and (b) a walled emitter structure used for the same purpose in III-V devices.

multiple values for the sheet resistance and appropriate geometry scaling are considered [SKLC08].

4.5.2.2 Walled emitter structure

Fig. 4.42a shows the top-view of a typical bipolar tetrode structure used in a SiGe process for the purpose of explaining the basic concept of its use. A simple walled-emitter (walled-E) structure was used in [Sch88] for measuring R_B at low bias. Since walled-E structures were difficult to fabricate in planar silicon BJT processes, a circular tetrode was employed in [RS91] along with a newly developed method for measuring R_B under all bias conditions. The method was later simplified and improved in [SL07].

An emitter ring around a base contact can often not be fabricated in an InP process. A walled-E structure like the one in fig. 4.42b serves the same purpose though, since the extension of the emitter to the edge of the base mesa prevents any current flow around the emitter window. This is in fact the most simple implementation of a tetrode.

Applying a voltage ΔV_{BB} between the separate base contacts B_1 and B_2 forces a lateral current I_{BB} to pass through the internal base region under the emitter. If ΔV_{BB} is kept sufficiently small (e.g., ≤ 10 mV), the current is just superimposed on the vertical DC base current flowing through the BE

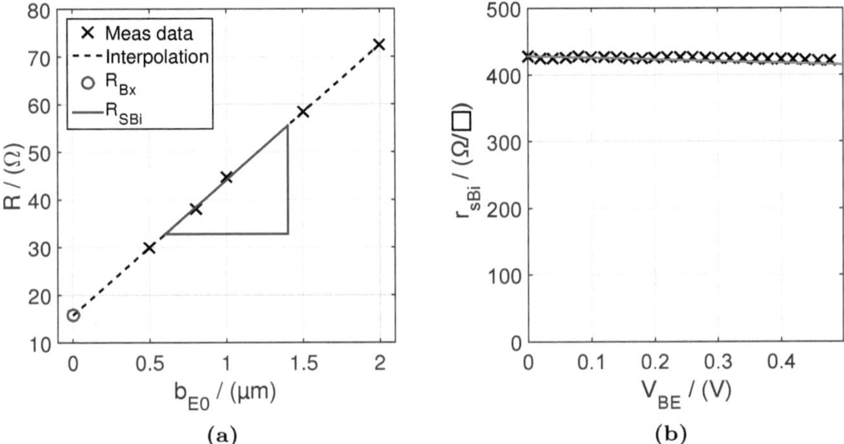

Figure 4.43: (a) Example for the determination of the internal base sheet resistance from the slope of the resistance for multiple measured tetrodes and (b) bias dependence of the internal base sheet resistance determined from the extraction according to (a) for multiple bias points and from the compact model (line).

diode (assuming a non-conducting BC diode, e.g. by setting $V_{BC} = 0$). All measurements are performed with grounded emitter.

Using two tetrode structures with different emitter lengths allows the elimination of any impact of the foreside and edges by subtracting the currents of the two structures. This leaves the current ΔI_{BB} which is equivalent to that of a two-dimensional (2D) structure with emitter length Δl_E. The resulting total measured resistance is then

$$R = \frac{\Delta V_{BB}}{\Delta I_{BB}} = 4R_{Bx} + r_{SBi}\frac{b_E}{\Delta l_E} \qquad (4.5.2\text{-}73)$$

and is linked directly to the external base resistance R_{Bx} of a two-base and one-emitter finger transistor as well as to the bias dependent sheet resistance r_{SBi} of the internal base region. By plotting the measured R vs. $b_E/\Delta l_E$, above equation allows to accurately determine R_{Bx} and r_{SBi}. An application example is shown in fig. 4.43a.

The rightmost expression of (4.5.2-73) is related to the lumped internal DC base resistance calculated from the device geometry as

$$R_{Bi} = r_{SBi}\frac{b_E}{l_E}g_i(b_E, l_E)\psi, \qquad (4.5.2\text{-}74)$$

where g_i is a geometry function ranging from $1/12$ for a transistor with two long double-base stripes to $1/28.6$ for a square-emitter transistor with surrounding base contact according to [SC10, Sch91a, SKLC08, YZ13]. ψ is the DC emitter current crowding function; its value equals 1 for sufficiently low forward bias and is very close to 1 even at higher bias in transistors used for high-frequency applications. The actual compact model parameter resulting from the tetrode measurement is the zero-bias internal base resistance $R_{\mathrm{Bi0}} = r_{\mathrm{sBi0}} b_{\mathrm{E}} g_i / l_{\mathrm{E}}$, since the bias dependence of r_{sBi} and ψ is calculated in a compact model. An example for the bias dependence of r_{sBi} is shown in fig. 4.43b.

The tetrode in combination with simple contact chain-type structures has been successfully applied for over thirty years to a large variety of technologies of many different foundries. Since most SiGe HBT PDKs offer scalable and statistical modeling, the results from these test structures have been routinely used to set up these models. In order to establish a similarly reliable extraction flow for III-V HBTs, the addition of the special test structures to a standard model parameter extraction environment is strongly recommended.

From the measurable bias dependence of the internal base sheet resistance, the zero-bias hole charge Q_{p0} can be determined if the junction capacitance parameters are known from

$$r_{\mathrm{sBi}} = r_{\mathrm{sBi0}} \frac{Q_{\mathrm{p0}}}{Q_{\mathrm{p0}} + Q_{\mathrm{jEi}} + Q_{\mathrm{jCi}}}. \tag{4.5.2-75}$$

A disadvantage of the test structure method is the required accurate knowledge about the actually fabricated transistor dimensions. Any difference between the assumed and actual dimensions of both the tetrode and the transistor structure results in errors in the calculated base resistance value. Also, ΔV_{BB} should not be chosen too large to avoid nonlinear effects.

4.5.2.3 Structure for the dynamic collector resistance

The structures for the dynamic, i.e. frequency dependent, collector resistance are not listed in tab. 4.7 since they were implemented on the test chip evaluated here for the first time. The purpose is to investigate the possible need for splitting the external collector resistance in the equivalent circuit into a constant and a frequency dependent fraction.

The structure is designed in analogy to the walled emitter structure with two separately biased collector contacts. It was placed in GSG pads since it was intended for small-signal measurements. Corresponding deembedding structures were also fabricated. The base and emitter contacts are both connected with the ground, i.e. $V_{\mathrm{BE}} = V_{\mathrm{ES}} = 0$. The schematic view of the structure is shown in fig. 4.44a.

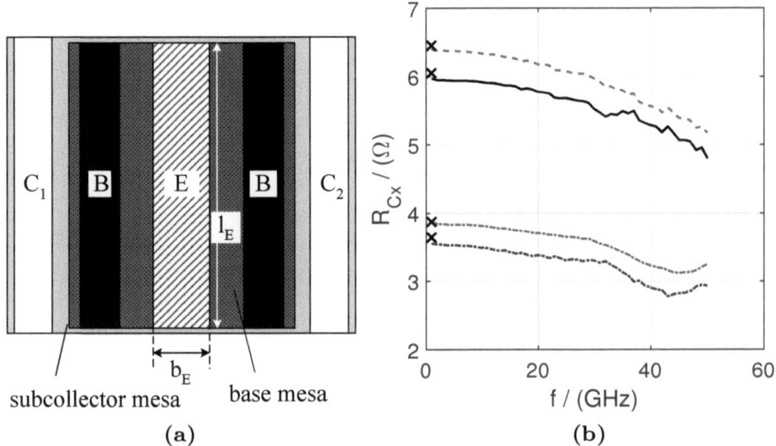

Figure 4.44: (a) Schematic view of the test structure for measuring the dynamic collector resistance and (b) measured collector resistance of four structures with different l_{E0} and b_{E0}. The symbols in (b) indicate the value for R_{Cx} obtained from DC measurements, showing good agreement at low frequencies.

The DC resistance of the structure can be evaluated by forcing the current through one contact to zero and measuring the voltage, thereby effectively measuring the voltage drop over the external collector resistance $V_{RCx} = V_{C1} - V_{C2}$. From small signal measurements, R_{Cx} is easily obtained as

$$R_{Cx} = \mathrm{Re}(\underline{Z}_{11} - \underline{Z}_{12}), \tag{4.5.2-76}$$

since each of the two collector contacts corresponds to one one of the ports. The resulting resistance is shown in fig. 4.44b.

Based on the test structure results, a first evaluation of the compact model with a frequency-dependent collector resistance showed a negligible impact on the model characteristics; the effect was therefore not implemented in the compact model.

4.5.3 Geometry separation

Many of the elements of a transistor in a compact model are assigned either to the internal or the external resp. peripheral region. Separating the two contributions is important not only because they may have a different bias dependence due to, e.g., a different doping profile, but also to model the frequency dependence correctly. The most common approach (e.g [Rei84,

SRR$^+$99]) is the separation into the emitter perimeter and area components, often called PA-separation. A transistor quantity M scales according to

$$M = \overline{M} A_{\mathrm{E0}} + M' P_{\mathrm{E0}}. \tag{4.5.3-77}$$

The parameters $\overline{M}$ and M' are separated by using a linear regression on M/A_{E0} w.r.t. $P_{\mathrm{E0}}/A_{\mathrm{E0}}$, where the slope gives the perimeter component M' and the intersection with the y-axis gives the area-dependent component $\overline{M}$. The separation is either performed for each bias point if separate parameter sets are to be extracted or for a single model parameter (e.g. the emitter diode saturation current I_{BEs}). The same approach can be used to separate the internal and external contributions of, e.g., the BC junction capacitance by using the external collector area A_{Cx} instead of the emitter perimeter.

An example for the separation is shown in fig. 4.45 for the BE junction capacitance, where the built-in voltage and the exponential coefficient were determined for the reference device and subsequently, C_{jE0} and C_{BEpar} were fitted for all devices with $l_{\mathrm{E0}} = 15\,\mu\mathrm{m}$. Fig. 4.45a shows the linear extrapolation, yielding a zero perimeter dependence for the zero-bias junction capacitance and only a weak perimeter dependence for the parasitic capacitance, which was subsequently set to zero. The resulting agreement between model and measurement over bias is shown in fig. 4.45b. It does not, however, perform similarly well when multiple device lengths are considered (cf. discussion in the next section).

An alternative to the PA-separation is bilinear scaling, as introduced in [Sch13] and discussed in [Paw14]. The quantity M is then scaled by

$$M = \overline{M} A + M_{\mathrm{b}}' b + M_{\mathrm{l}}' l + M_{\mathrm{C}}, \tag{4.5.3-78}$$

with the area of the appropriate region $A = b \cdot l$, its width b and its length l. M_{C} is a constant corner component. The PA scaling is a special case of the more general bilinear scaling in which $M_{\mathrm{C}} = 0$, and $M_{\mathrm{b}}' = M_{\mathrm{l}}' = 2M'$. The application and physical interpretation are discussed in [Paw14].

4.5.3.1 Problems with the approach

Geometry separation for parameter determination and of models in general only works for scalable processes, i.e. the drawn and actual device dimensions must be related in a known and reproducible way. For SiGe devices, scalable modeling has been practiced for decades. Process variations are routinely investigated and are well known. For III-V devices, on the other hand, a limited amount of discrete transistors (typically three to five) are offered by a foundry for which individual models based on single-transistor fitting are

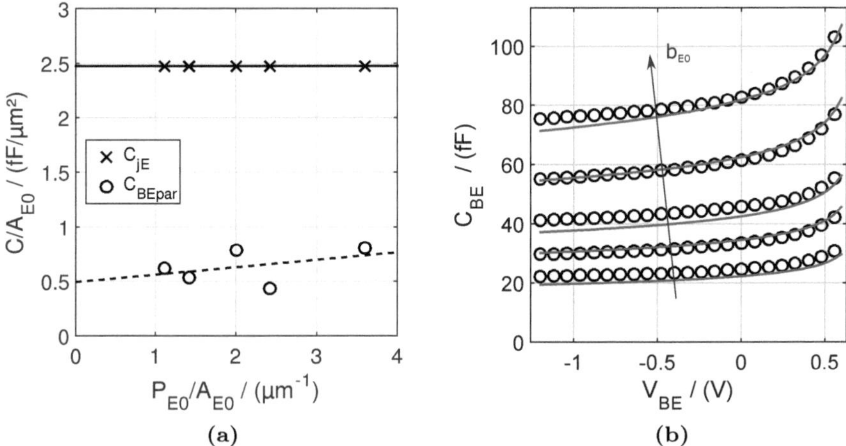

Figure 4.45: (a) C_{jE0} and C_{BEpar} for all devices with $l_{\mathrm{E0}} = 15\,\mathrm{\mu m}$ (symbols) along with the linear regression (lines) and (b) comparison of model (lines) and measurement (symbols) for all devices with $l_{\mathrm{E0}} = 15\,\mathrm{\mu m}$ using the scaling determined in (a) and neglecting the perimeter component of C_{BEpar}.

supplied. Therefore, process scalability is less of a focus for these processes than individual device performance.

Investigation of process variations is a time-consuming and above all expensive process due to the requirement of specialized equipment for, e.g., focused ion beam (FIB) cutting and scanning electron microscopy (SEM). Even though III-V devices are comparatively easy to analyze due to the mesa structure and the clear separation of semiconductor material from the surrounding passivation, only a few such pictures could be taken over the course of this work, investigating the cross-section along the length of the device with $(l_{\mathrm{E0}}, b_{\mathrm{E0}}) = (15,0.8)\,\mathrm{\mu m}$ and along the width of the device with $(l_{\mathrm{E0}}, b_{\mathrm{E0}}) = (15,0.5)\,\mathrm{\mu m}$. Different devices were used because the FIB cut destroys the device. Examples for the obtained pictures are shown in fig. 2.4. Tab. 4.9 lists a comparison of the drawn measurements with those obtained from the SEM pictures; for the definition of the geometry parameters, compare fig. 2.3.

Additional clues with regard to differences between drawn and actual transistor dimensions are provided by the application of extraction methods to the terminal data of the devices. The extracted collector resistance of the GCS devices with $l_{\mathrm{E0}} = 15\,\mathrm{\mu m}$ decreases with increasing b_{E0} (cf. section 4.3.4.5). This

Table 4.9: Comparison of device dimensions drawn in the layout and measured by SEM/TEM evaluation.

Param.	Drawn dim.	Measured dim.	Comment
l_{E0}	15 µm	13 µm	Emitter mesa length at BE junction
l_{bc}	0.5 µm	0.14 µm	
l_{bl}	0.5 µm	0.85 µm	
l_{KC}	2.8 µm	1.95 µm	
l_{BE}	0.6 µm	0.61 µm	
l_{ldg}	0.2 µm	0.33 µm	Distance between emitter metal and base metal towards base via
b_{E0}	0.5 µm	0.58 µm	Emitter mesa length at BE junction
b_{KB}	0.6 µm	0.5 - 0.55 µm	asymmetrical
b_{KC}	2.3 µm	2.0 µm	
d_{bc}	0.65 µm	0.7 - 0.8 µm	asymmetrical

implies increasing length, since the external collector resistance scales roughly with b_{E0}/l_{E0}. However, the extraction methods are insufficiently precise to determine the actual device dimensions and can merely indicate deviations.

As a consequence of this first investigation, the test structure according to fig. 4.46 is proposed here. A multi-finger transistor with different b_{E0} and a constant reference l_{E0} for each finger and a second multi-finger transistor with a different l_{E0} and a constant reference b_{E0} without contacts are used for determining the actual transistor dimensions. FIB cuts are required once per structure perpendicular to the varied dimension; SEM pictures are taken at multiple points during the cut. Since the transistors are not intended for measurement, no contact pads need to be provided and the structure can be placed at a nearly arbitrary location, saving wafer space. So far, however, the structure has not been implemented.

For this work, based on the analysis of the SEM pictures, the effective emitter window dimensions

$$l_{E0} = l_{E0,\text{drawn}} - 2\,\mu\text{m} \quad \text{and} \quad b_{E0} = b_{E0,\text{drawn}} + 0.08\,\mu\text{m} \qquad (4.5.3\text{-}79)$$

were used. Additionally, effective dimensions were determined by applying PA scaling with a concurrent nonlinear fit of Δb_E and Δl_E in order to obtain the smallest possible average deviation. If extracted in this manner, the effective dimensions are $l_{E0} = l_{E0,drawn} + 1.17\,\mu m$ and $b_{E0} = b_{E0,drawn} + 0.08\,\mu m$, which is quite close to those obtained from the SEM pictures.

Then, the best obtainable accuracy using bilinear scaling is shown in fig. 4.47 under the assumption that the compact model can reproduce the components of the scaling equation exactly for every bias point. Realistic agreement between the final compact model and the measurement can be worse. Fig. 4.47 was included to show that, even under ideal circumstances, relative errors of $> 25\%$ must be expected for some transistor sizes, though for most devices, the error is $< 10\%$.

Additional problems may occur when a quantity shows an extreme deviation from the expected behavior, i.e. when some parasitic or unexpected effect covers the actual scalable parameter. For the investigated technology, the base-collector current at $V_{BE} = 0\,V$ proved to have a very large saturation current, but showed a resistive component at medium V_{BC} before returning to the expected exponential behavior, i.e. a diode in series with a large resistance was included in parallel with the actual junction current (cf. fig. 4.48, which also demonstrates that scalability of the BC current is not very good). Furthermore, the saturation current showed a negative internal area component when using (4.5.3-77) for separating the A_{E0} and A_{Cx} dependence. This effect is atypical for the investigated technology, since other devices from a previous process run did not show the same characteristic. While it was included in the model using an additional diode and resistance, it was not investigated at length due to (i) the unavailability of simulations as discussed in section 2.4 and (ii) the effect being unintentional and not present in future technology versions. A potential source is a bad passivation between the base and collector contacts, permitting a limited current flow.

4.5.4 Scaling of other parameters

While the resistances are scaled according to the known geometry dependence according to, e.g., [SKLC08] for the base and [RKP$^+$07] for the collector, and the internal and external elements follow the simple separation equation (4.5.3-77) or a similar approach, e.g. the bilinear scaling (4.5.3-78), several other compact model elements have not been discussed so far.

Most importantly, the emitter resistance consists of a constant contribution from the metalization and an inversely emitter-area dependent contribution from the transistor emitter cap layer. It scales approximately according to

$$R_E = R_{Ec} + \frac{R_{Ea}}{A_{E0}}, \tag{4.5.4-80}$$

with an example comparing the result of the R_E value as determined by extraction methods and the model scaling equation given in fig. 4.49a. The

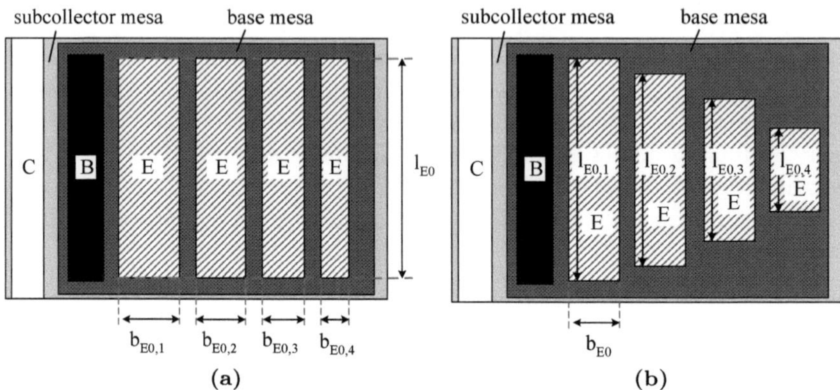

Figure 4.46: Proposed SEM test structure with (a) different b_{E0} and (b) different l_{E0}. The emitter window number is arbitrary and depends on the technology.

agreement is very good, especially considering that extraction methods may have a geometry-dependent error term (cf. section 4.3).

The scaling of the thermal resistance can become very complicated and may require thermal simulations to fully understand [Zim16]. The latter require a well-known device layer structure, since different semiconductor materials can have widely differing thermal conductivities. For example, InGaAs, commonly used in the base layer of InP HBTs, is an excellent thermal insulator, providing a high thermal resistance. Here, very good agreement between the extracted R_{th} for each device individually and model scaling was obtained by using

$$R_{th} = \frac{R_{th0}}{1 + a_{thb}b_{E0} + a_{thl}l_{E0}}.$$ (4.5.4-81)

The comparison is shown in fig. 4.49b, along with a thermal simulation of the devices placing the thermal sink at the bottom of the device.

Additionally, the transit times were found to be independent of the device size, with only r_{Ci0} scaling in analogy to (4.5.4-80) and the parasitic BE capacitance was found to scale reasonably well with the emitter width.

4.5.5 Discussion

In this section, the extraction of scalable model parameters for an InP technology was demonstrated based on examples for each step. Problems encountered during the process were discussed, and a potential solution by using a new test

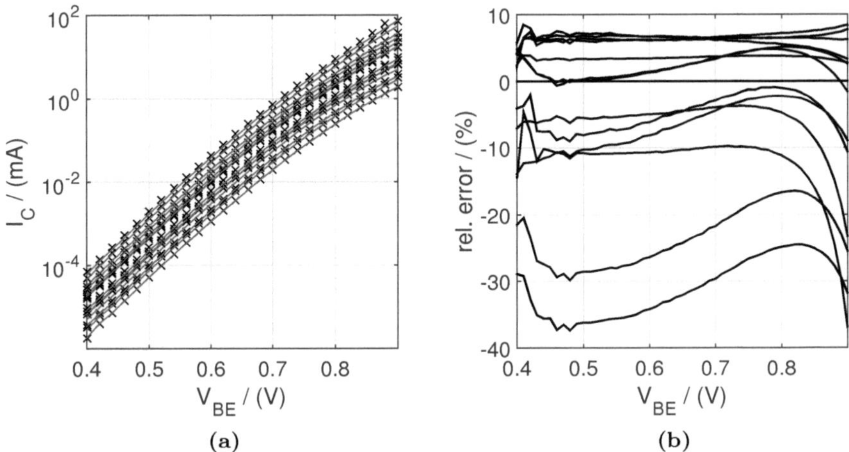

Figure 4.47: (a) Comparison of bilinear scaling of the collector current at $V_{BC} = 0\,V$ for all available devices using the effective emitter dimensions (4.5.3-79) for each bias point (symbols: measurement, lines: bilinear scaling) and (b) relative error of the scaling equations w.r.t. measurement.

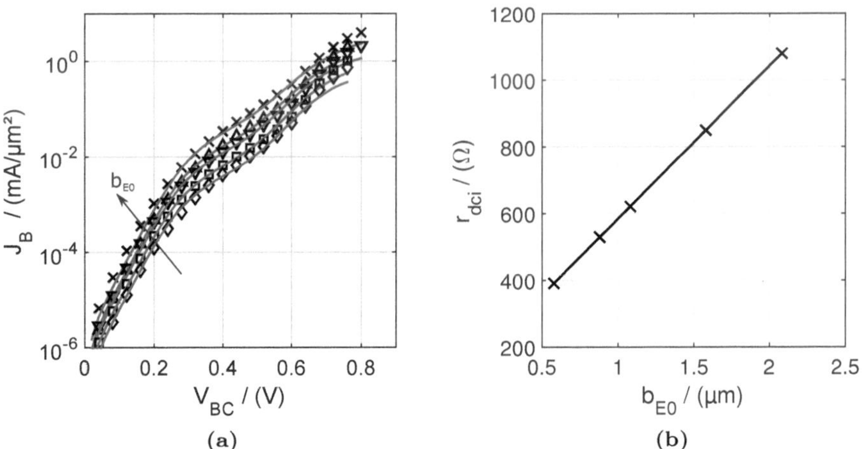

Figure 4.48: (a) Base current for $V_{BE} = 0\,V$ from measurements (symbols) and model (lines); (b) extracted series resistance for the BC diode, both for devices with $l_{E0} = 15\,\mu m$

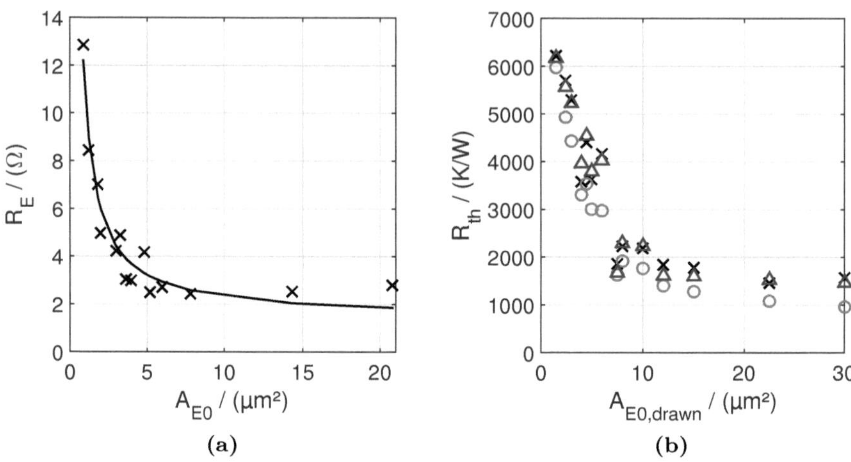

Figure 4.49: (a) Comparison of extracted (symbols) and modeled (line) emitter resistance according to (4.5.4-81). (b) Comparison of R_{th} extracted for each device (crosses) with scaling equation (4.5.4-81) (triangles) and thermal simulation (circles). Shown w.r.t. the drawn emitter area since the effective area was unknown for the simulations.

structure sketched out. It was also shown that many principles of compact modeling commonly used in SiGe processes, such as using the tetrode for the extraction of the bias-dependent base resistance, are also applicable to III-V technologies. A comparison of the scalable model with measurements for the process discussed in this section is shown in section 5.1.

Overall, a scalable model for III-V devices can, at this point, not compare in maturity or performance with a scalable model for a SiGe process. Given that in the latter case usually multiple process generations are modeled with iterative improvements and close communication with process engineers that helps clear up deviations between measurement and model, matching SiGe model quality in III-V is also not expected at this point. However, considering the difficulties of creating the first scalable model for a given process, the obtained agreement is very good. Further cooperation between model and process engineers is required to improve the performance as both the knowledge about the process and its stability grow.

While the model shows acceptable scalability for most devices and model cards for multiple lengths and widths were delivered to the foundry, the focus for the parameter extraction and scaling was placed on the long devices as those closest to the ideal case (i.e. relatively low foreside impact).

In conclusion, while a scalable model performance can never be better than the process repeatability dictates, good agreement for a wide range of devices was reached in this study.

<h1>CHAPTER 5</h1>

Model application

This chapter demonstrates the application of the methods described in this work to three different III-V processes: the GaAs technology that was used as the example in the description of the single-transistor extraction in section 4.4, the InP HBT technology with $(f_\mathrm{T}, f_\mathrm{max}) = (300, 250)$ GHz [GCS] that was used in the examples for the geometry-scalable extraction in section 4.5 and a high-speed InP technology with $(f_\mathrm{T}, f_\mathrm{max}) = (350, 500)$ GHz [GDR$^+$05] which demonstrates the need for the multi-region C_BC and current-dependent τ_c model.

The results in this chapter are presented for $T = 300\,\mathrm{K}$ unless mentioned otherwise. For all small-signal measurements, open-short deembedding was performed.

5.1 250 GHz InP technology

A fully scalable model for the InP HBTs manufactured in this technology was created and results were published in [NSC$^+$13]. Though the model scales quite well for both emitter length and emitter width, the focus for the application was put on the width scaling for very long transistors, i.e. $l_\mathrm{E0} = 15\,\mu\mathrm{m}$, representing a nearly ideal case with little impact of the transistor foreside, allowing for an easier parameter extraction.

The agreement of the measured collector current with the model is shown in fig. 5.1 for both width- and length-scaling. Excellent agreement is obtained for both.

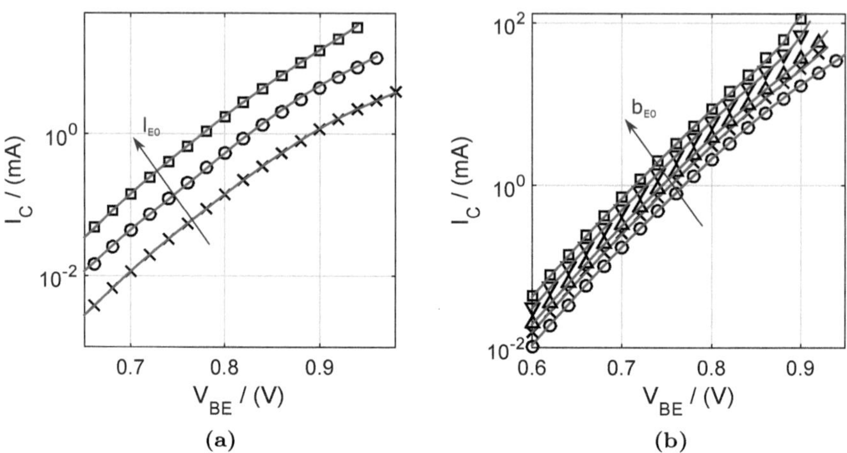

Figure 5.1: Comparison of the model (lines) with measurements (symbols) for $V_{BC} = 0\,V$, $T = 300\,K$ and (a) $b_{E0} = 0.8\,\mu m$, $l_{E0} = [3, 5, 10]\,\mu m$ and (b) $l_{E0} = 15\,\mu m$, $b_{E0} = [0.5, 0.8, 1, 1.5, 2]\,\mu m$

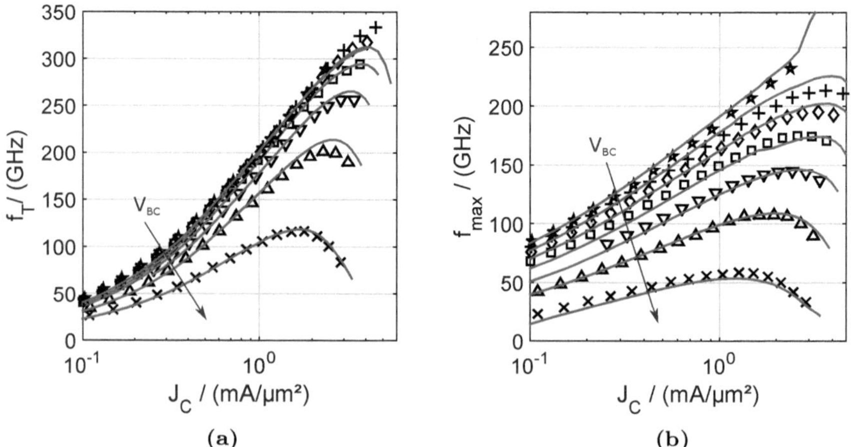

Figure 5.2: Comparison of the model (lines) with measurements (symbols) the reference device with $(b_{E0}, l_{E0}) = (0.8, 15)\,\mu m$ for $V_{BC} = [-0.25, 0, 0.1, 0.2, 0.3, 0.4, 0.5]\,V$ and $T = 300\,K$. (a) f_T and (b) f_{max}. The sharp increase in (b) for high current densities is due to thermal instability of the model outside the measurable range.

The model parameters for the charges are verified by the transit and oscillation frequencies. Both are displayed for the reference device, i.e. the device for which the non-scaling parameters were determined, in fig. 5.2. The transit frequency was obtained from

$$f_\mathrm{T} = \frac{f_\mathrm{meas}}{\mathrm{Im}\left(\frac{\underline{Y_{11}}}{\underline{Y_{22}}}\right)},\tag{5.1.0-1}$$

while f_max was found by fitting a first order low-pass to the unilateral gain for each bias point, since using a spot frequency was found to produce noisy data. The scaling of the small-signal parameters is verified by the plots of the transit frequency for multiple devices in fig. 5.3.

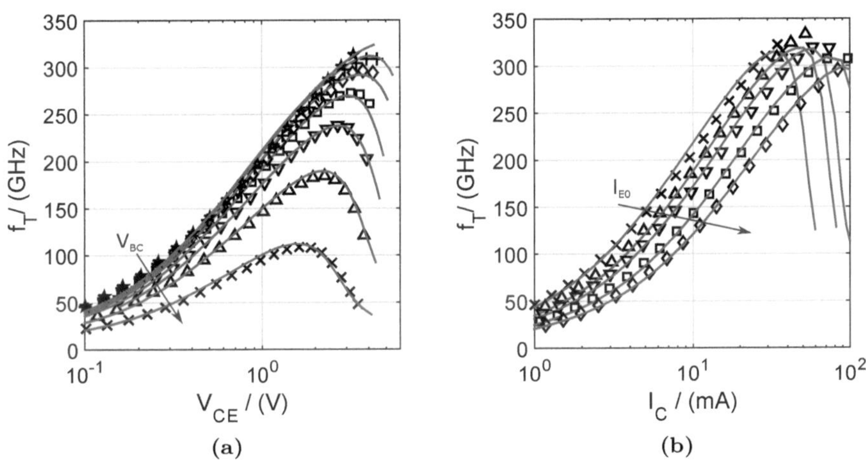

Figure 5.3: Comparison of the model (lines) with measurements (symbols). (a) f_T for the device with $(l_\mathrm{E0}, b_\mathrm{E0}) = (3, 0.8)$ µm at $V_\mathrm{BC} = [-0.25, 0, 0.1, 0.2, 0.3, 0.4, 0.5]$ V. (b) f_T for the devices with $l_\mathrm{E0} = 15$ µm and $b_\mathrm{E0} = [0.5, 0.8, 1.0, 1.5, 2.0]$ µm at $V_\mathrm{BC} = 0$ V. Both for $T = 300$ K.

Further verifying the model performance for the reference device, the collector current for a constant BE voltage, i.e. the forward output characteristic, is shown in fig. 5.4a. For comparison, since III-V models in practical use still do not always include the effect [Sak16], the model without self-heating is shown, clearly demonstrating a larger deviation from the measured data. The maximum available gain (MAG) shown in fig. 5.4b is also modeled well.

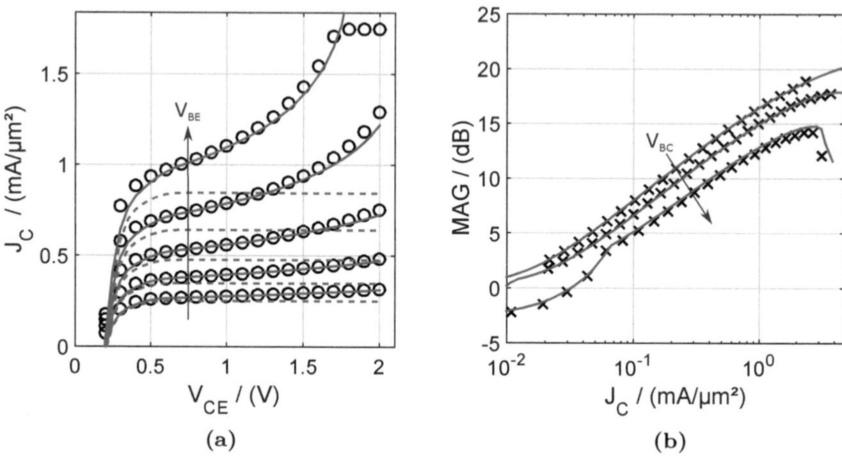

Figure 5.4: Comparison of the model with (solid lines) and without (dashed lines) self-heating with measurements (symbols) for the reference device with $(b_{E0}, l_{E0}) = (0.8, 15)$ µm. (a) Forward output characteristic for $V_{BE} = [0.8, 0.815, 0.83, 0.845, 0.86]$ V. The dashed line shows the model without self-heating. (b) MAG for $f = 20\,$GHz and $V_{BC} = [-0.25, 0.2, 0.4]$V. Both for $T = 300\,$K.

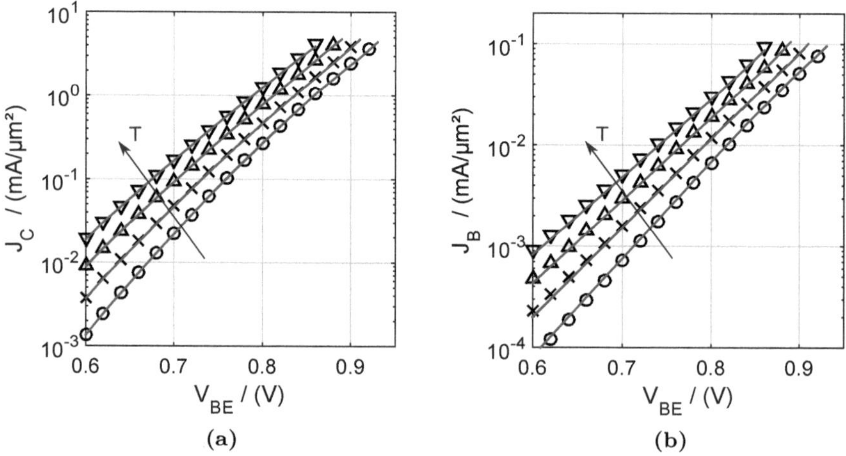

Figure 5.5: Comparison of the model (lines) with measurements (symbols) for the reference device with $(b_{E0}, l_{E0}) = (0.8, 15)$ µm. $V_{BC} = 0\,$V and $T = [300, 325, 350, 375]$ K. (a) J_C and (b) J_B.

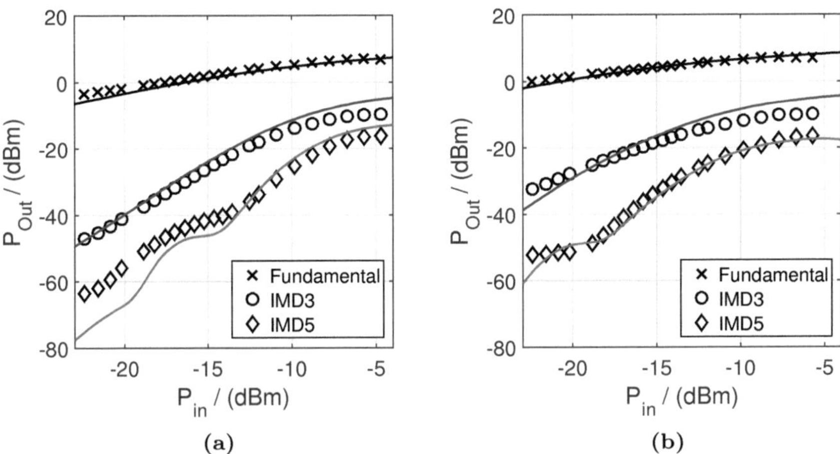

Figure 5.6: Comparison of the model (lines) with measurements (symbols). Output power at fundamental frequency, third and fifth harmonic frequencies for $V_{CE} = 1\,V$, $f_0 = 10\,GHz$, $T = 300\,K$ and (a) $V_{BE} = 0.82\,V$, (b) $V_{BE} = 0.86\,V$.

Fig. 5.5 demonstrates the modeling of the currents for different ambient temperatures. Finally, the large-signal agreement of the model was investigated for typical operating points with results shown in fig. 5.6. It should be noted that no input and output currents and voltages measured during the power sweeps were available; the exact load and source resistances could therefore not be obtained as proposed in, e.g., [Paw14]. This is the likely cause for the deviations seen in fig. 5.6. Additional plots for this technology can be found in appendix C.1.

5.2   500 GHz InP technology

A scalable model was extracted for the InP HBT technology with $(f_\mathrm{T}, f_\mathrm{max}) = (350, 500)$ GHz. However, while the model for the reference device with $(l_{E0}, b_{E0}) = (4, 0.25)$ µm agrees very well with measurements, scalability is not good for this technology. No SEM or TEM pictures investigating the actual compared to the drawn device dimensions were available. Transistors of different width demonstrate slightly different behavior over bias, potentially caused by bias-dependent scaling, which has not been investigated at the time of this writing. Additionally, reproducibility is not sufficient; this is demonstrated in fig. 5.7a. Acceptable results are obtained for $b_{E0} \leq 0.35$ µm and $l_{E0} \leq 6$ µm. The available devices are listed in tab. 5.1.

Fig. 5.7b shows the BC junction capacitance for the reference device at multiple temperatures, determined from the Y-parameters as described in section 4.4.1, demonstrating the need for the multi-region C_BC model. The agreement is acceptable, but could be improved by an individual fit to the reference device; the deviations visible in the figure are owed to the model scalability.

The forward gummel characteristic for the reference device is shown in fig. 5.8. Fig. 5.9 gives the comparison for f_T and f_max. The former shows the crossover typical for the NDM effect in the collector as discussed in section 3.5.

An impression for the scalability of the model is given by the comparison of the forward output characteristic in fig. 5.10 for the reference device and a second transistor with $(l_{E0}, b_{E0}) = (6, 0.2)$µm.

Finally, while no large-signal data are available for this technology, noise parameters were measured and published in [SNS+16]. An example is shown in fig. 5.11. The agreement is very good for medium to high current densities. The deviation between model and measurement for low current densities also

Table 5.1: Available transistors in GSG pads for the 500 GHz InP technology.

			$l_{E0}/\mu m$			
$b_{E0}/\mu m$	2	4	6	8	10	15
0.2		x	x			
0.25	x	x		x	x	x
0.35		x				
0.5		x		x		
0.7	x	x				

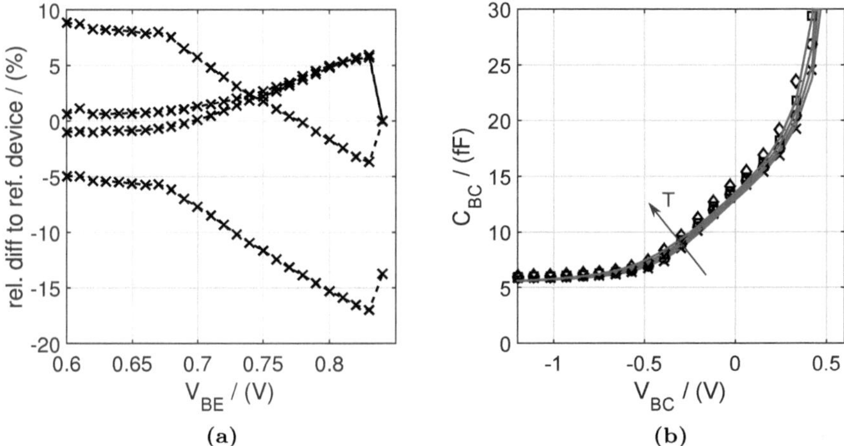

Figure 5.7: (a) Relative difference of multiple devices with drawn emitter dimensions, from the same wafer and different cells, w.r.t. the modeled device. (b) Measured (symbols) and modeled (lines) BC junction capacitance determined from cold S-parameter measurements for the reference device with $(b_{E0}, l_{E0}) = (0.25, 4)$ µm and $T = [300, 325, 350, 375]$ K.

occurs for SiGe devices and is not subject of the investigations in this work. Additional plots for the technology are shown in appendix C.2.

Despite the problems with the model scaling, the figures in this section demonstrate very good agreement of the HICUM/L2 compact model with the measured data. Simultaneously, they show that the model extensions as discussed in chapter 3 are necessary for accurately representing modern, high-speed technologies.

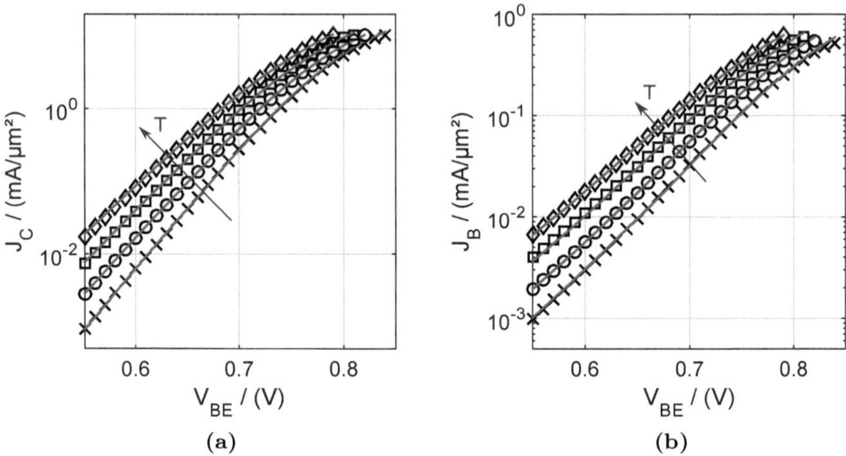

Figure 5.8: Measured (symbols) and modeled (lines) currents. (a) J_C and (b) J_B for the reference device with $(b_{E0}, l_{E0}) = (0.25, 4)$ µm, $T = [300, 325, 350, 375]$ K and $V_{BC} = 0$ V.

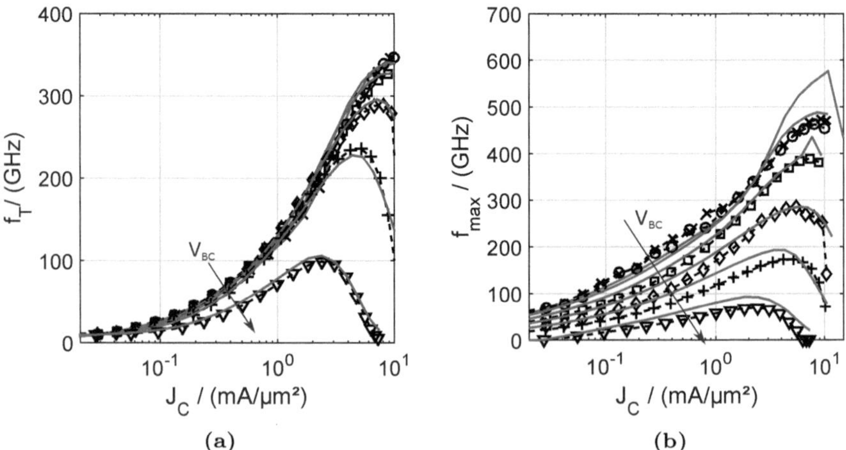

Figure 5.9: (a) f_T and (b) f_{max} for the reference device with $(b_{E0}, l_{E0}) = (0.25, 4)$ µm, $T = 300$ K and $V_{BC} = [-0.75, -0.5, -0.25, 0, 0.25, 0.5]$ V. The increase in f_{max} for $V_{BC} = -0.75$ V is caused by the NDM model and still under investigation at this point, since the compact model implementation for $\Delta\tau_{BCv}$ is preliminary. Symbols: measurement, lines: model.

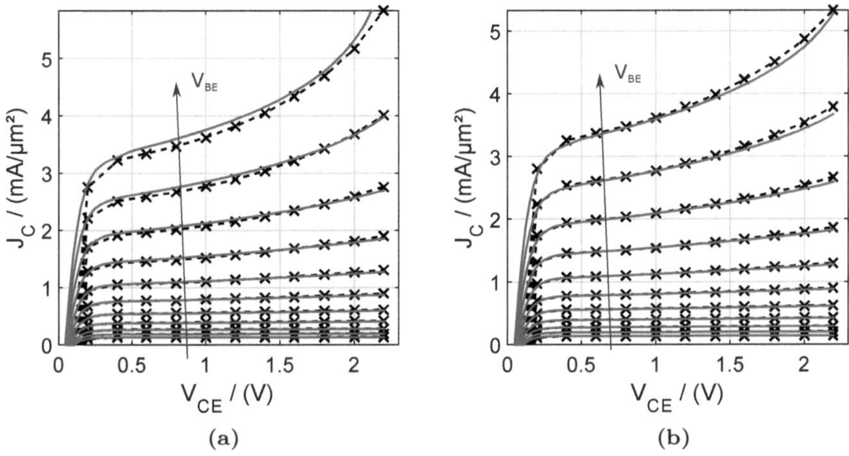

Figure 5.10: Measured (dashed lines with symbols) and modeled (solid lines) J_C for $V_{BE} = 0.68 \ldots 0.78$ V and $T = 300$ K, (a) $(l_{E0}, b_{E0}) = (4, 0.25)$ µm and (b) $(l_{E0}, b_{E0}) = (6, 0.2)$ µm

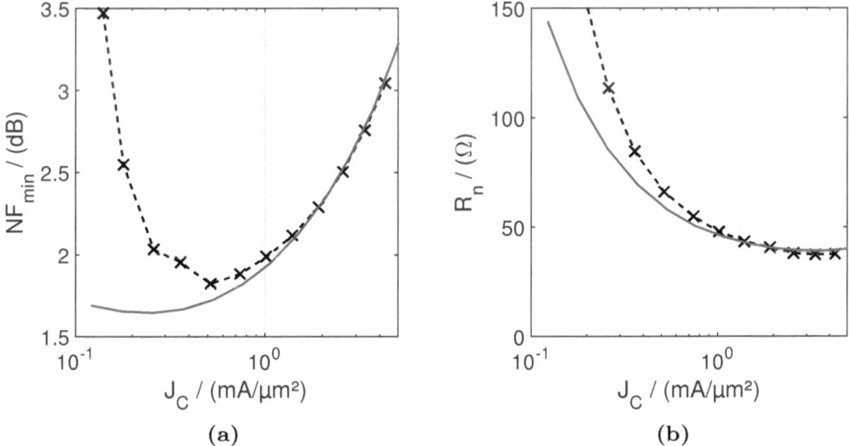

Figure 5.11: Measured (dashed lines with symbols) and modeled (solid lines) noise parameters for the reference device with $(b_{E0}, l_{E0}) = (0.25, 4)$ µm, $T = 300$ K, $V_{CE} = 1.2$ V and $f = 10$ GHz (a) minimum noise figure NF_{min} and (b) noise resistance R_n.

A dependence of the base current on the CE voltage during output characteristic measurements of this HBT technology was observed. An example is shown in fig. 5.12a. Increasing collector current and decreasing base current with increasing CE voltage is typical for the avalanche effect. However, the avalanche effect causes carrier multiplication and therefore depends on the amount of injected carriers, i.e. the transfer current. The measured effect depends only on the BC voltage.

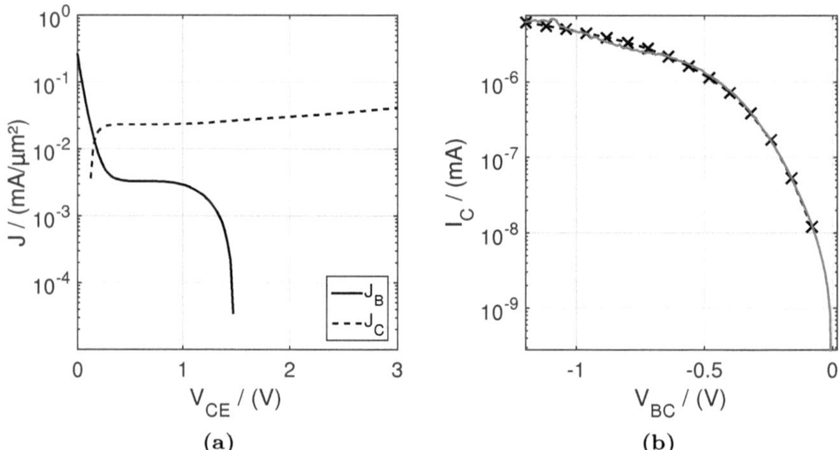

Figure 5.12: (a) Base and collector current densities of affected HBT (reference device, $T = 300\,\mathrm{K}$) during forward output measurement with constant V_{BE} and (b) comparison of model formulation (line) with measurement (symbols).

Two physical effects are possible as the origin of the current: tunneling of carriers through the reverse-biased BC junction and an imperfect isolation between the base and collector regions causing a leakage current. From the measured behavior, it is impossible to distinguish between them, though communications with the manufacturer suggest the latter [Urt16].

The current was included as a series connection of a diode and a resistance between the external base and collector node, with the diode saturation current I_{bcts}, the nonideality factor m_{bct} and the resistance R_{bct} as model parameters to be extracted from the $I_{\mathrm{C}} - V_{\mathrm{BC}}$-characteristic for a reverse-biased junction, resulting in the equation set

$$I_{\mathrm{bct}} = I_{\mathrm{bcts}} \left[\exp\left(\frac{V_{\mathrm{bct}}}{m_{\mathrm{bct}} V_{\mathrm{T}}} \right) - 1 \right] \text{ and} \qquad (5.2.0\text{-}2)$$

$$V_{\mathrm{bct}} = V_{\mathrm{BC'}} - I_{\mathrm{bct}} R_{\mathrm{bct}}. \qquad (5.2.0\text{-}3)$$

The comparison of the model with measured data is shown in fig. 5.12b. The agreement is very good. The jitter in the model curve is due to the fact that (5.2.0-2) and (5.2.0-3) are not explicitly solvable and the measured current is fed back into the model for the determination of the junction voltage. The relatively easy inclusion of this effect in the model demonstrates the advantage of model availability as VA code.

5.3 GaAs technology

The GaAs technology is included here as an example of the application of the compact model to a high-voltage technology, as opposed to the high-speed InP HBTs. However, since the measurements were not taken over the course of this work, only a limited voltage range is available for comparison. Since only a single device was measured and modeled, all plots are for a device with $(b_{E0}, l_{E0}) = (2.2, 22)$ µm.

The most important DC characteristics are shown in 5.13 for the forward gummel plot at multiple ambient temperatures and the forward output characteristic in fig. 5.14. Good agreement between model and small-signal measurements is demonstrated for f_T and f_{max} in fig. 5.15. Results regarding noise measurements of this technology using the compact model obtained here were also published in [SNS+16].

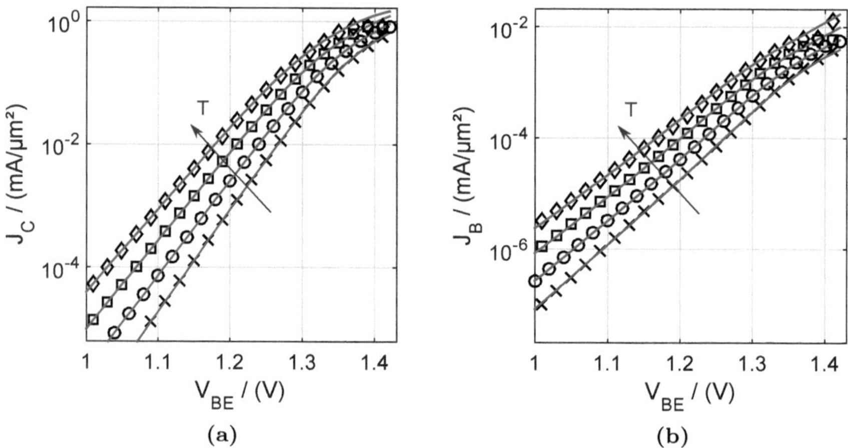

Figure 5.13: (a) J_C and (b) J_B, $T = [300, 325, 350, 375]$ K and $V_{BC} = 0$ V. Note, that the current was limited by the compliance setting during the measurement. Symbols: measurement, lines: model.

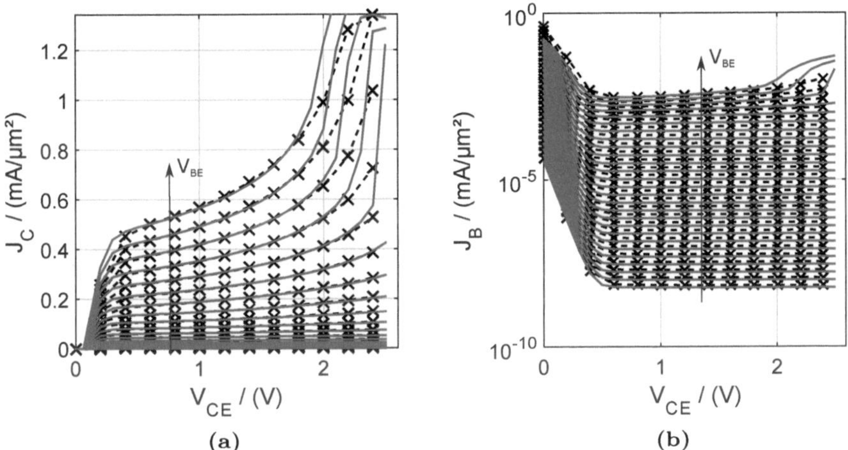

Figure 5.14: (a) J_C and (b) J_B for $V_{BE} = 0.9\ldots1.42$ V with a step width of $10\,\mathrm{mV}$ and $T = 300\,\mathrm{K}$. Overall, good agreement is demonstrated, although thermal breakdown is somewhat exaggerated in the model. Symbols: measurement, lines: model.

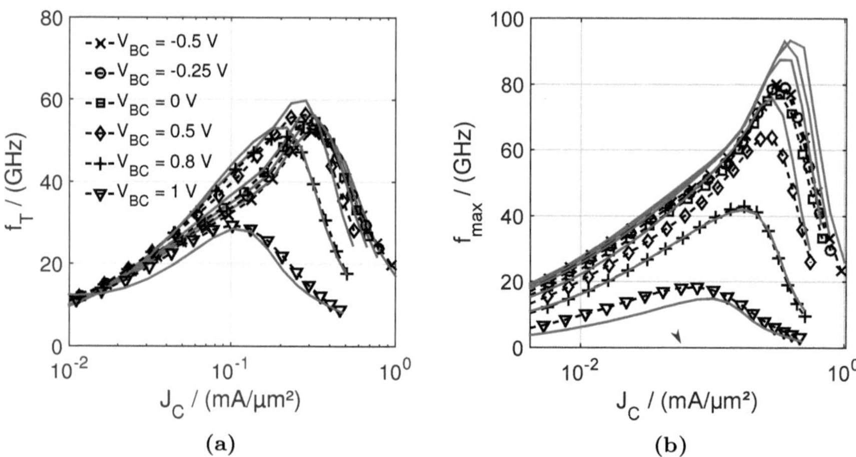

Figure 5.15: (a) f_T and (b) f_max for $V_{\mathrm{BC}} = [-0.5, -0.25, 0, 0.5, 0.8, 1]$ V and $T = 300\,\mathrm{K}$. The crossover of the transit frequency is clearly visible, with larger negative V_{BC} exhibiting lower f_T at medium current densities and higher f_T near the peak. The legend in (a) is valid for both plots. Solid lines correspond to the compact model. The deviation in f_max is likely caused by a too low value for the base resistance due to single-transistor extraction.

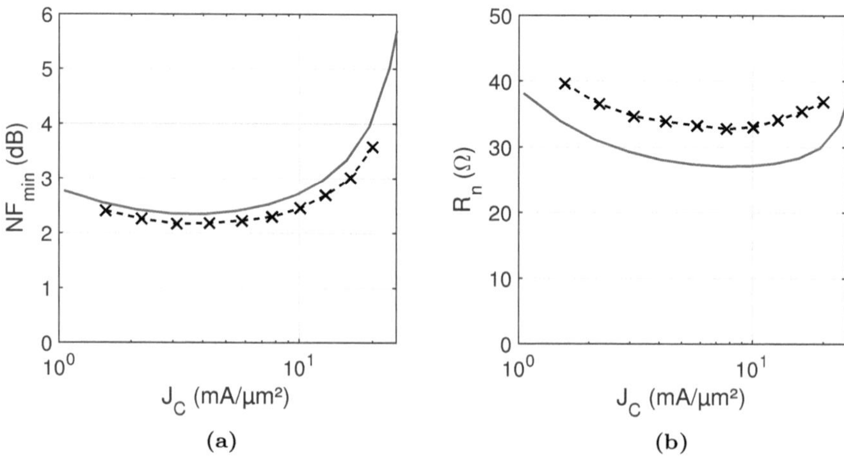

Figure 5.16: (a) Minimum noise figure and (b) noise resistance at $T = 300\,\mathrm{K}$, $f = 10\,\mathrm{GHz}$ and $V_{\mathrm{CE}} = 1.5\,\mathrm{V}$.

CHAPTER **6**

Summary and outlook

Over the course of this work, compact modeling capabilities for III-V HBTs were investigated and extended. In particular, the applicability of many standard methods for scaling and parameter extraction used commonly for SiGe HBT models, for which scalable compact modeling concurrent with process design has been common practice for years, was investigated. While III-V HBTs outperform silicon-based devices on many FOMs, these superior properties cannot be fully exploited without the appropriate infrastructure; this infrastructures includes fully scalable and above all accurate compact models, and a more systematic approach to both process improvements and modeling is urgently required.

First, the most important differences between SiGe HBTs and III-V HBTs were identified based on both the device structure and the physical effects relevant for both materials. With HICUM/L2, a compact model was chosen as a basis for this work that already contains many important formulations for physics-based device modeling for III-V materials, such as the separation of the large-area BC diode into an internal and external part, a bias-dependent reverse early effect and the current blocking effect.

The effects that were not included in the model already, most importantly the voltage- and current-dependent collector transit time behavior due to the NDM effect and the doping spike impact on the BC junction capacitance, were investigated in detail from semiconductor theory and included in the compact model. The use of numerical simulations for the investigation of current-dependent issues was considered but discarded, since standard simulators cannot model the NDM effect.

Since the accuracy of a compact model is always determined by the parameter extraction, a complete HICUM/L2 parameter extraction procedure was applied for both high-voltage and high-speed processes. The most common case for III-V modeling is the extraction of parameters for a single device, or a low number of devices with no test structures. Therefore, the focus of the applied extraction was on this situation, with particular care taken in the investigation of extraction methods for the series resistances as one of the earliest steps in the sequence with the most potential for errors affecting subsequent parameters. Many of the methods that are commonly in use in the industry were identified as insufficiently precise, causing large errors in the extracted resistances.

All required extraction steps for single-transistor and scalable parameter extraction were demonstrated based on III-V HBT measurement examples. It was also shown that self-heating, not the avalanche effect, is responsible for the increase of the collector current with increasing V_{CE} and that the large bandgap in the collector of III-V HBTs indeed suppresses avalanche carrier generation in the measurement range available in this work.

Geometry-scalable parameter extraction using a set of test structures and knowledge about the device geometry was discussed. A fully scalable model for an InP HBT process was created. Based on an investigation of the technology using FIB cutting and SEM pictures, scalability problems of the process were investigated; a new test structure for easier investigations in the future was proposed. Despite deviations between the drawn and actual transistor dimensions, good agreement between the scaled compact model and measurements was obtained.

A comparison of the model with measured data was shown for a variety of technologies and measurements. Overall, very good agreement between modeled and measured characteristics was demonstrated.

From the data presented here, several conclusions are possible. The most important is that physics-based, fully scalable compact modeling for III-V HBTs is possible if the appropriate effort in terms of test structure design, evaluation and wafer space is made and the important physical effects are included in the model. Then, the compact model data can predict circuit performance with a high degree of accuracy and can be used for further process improvements. Secondly, it has been shown that a model made for SiGe HBTs alone cannot capture all effects relevant for III-V HBTs and must be extended but also that a well-designed compact model with publicly available code forms an excellent basis for just these extensions.

From the scaling of the model, it is apparent that some current III-V processes need to be improved or at least investigated at length regarding the relation between the device dimensions as drawn in the layout and the ones after fabrication; this is a minor issue for single-transistor models that are common in the industry, but has a major impact if circuit designers have the freedom to insert arbitrarily sized HBTs into their designs.

While this work represents a major step towards systematic, consistent modeling of III-V devices it is by no means the last one. The split of both the external base and external collector resistance into two resistances each in order to model the distributed base-collector diode more precisely could not be investigated here, since the scaling of all process parameters needs to be known in detail for the separation of effects. So far, no iterative evaluation of the models, i.e. creation of models based on test structures, use under practical circuit condition and model improvement based on circuit performance feedback, was possible, but is desirable for further model development in order to better evaluate the model performance.

Improved investigation of the device behavior would be possible if reliable simulations for III-V devices can be obtained. The only possibility for this are BTE simulations. Material models and a boundary condition for the launcher in the conduction band of the BE junction are needed for accurate simulations, along with SIMS measurements of the doping and material profile. Among the open questions that could be answered by such simulations are the accuracy of the GICCR for III-V devices with very abrupt material transitions and the importance of the weight factors for the transfer current.

The creation of a full process design kit for any given technology entails a large amount of work; any process has to be investigated with regards to scaling irregularities and potentially be improved to obtain the full benefits of scalable compact modeling. Further, a full set of test structures has to be created, measured and may have to be iteratively improved. The created models need to be verified not only based on individual device but also on circuit measurements. If necessary for a given technology, scaling rules for multi-finger devices and other shapes, such as horseshoe emitters, need to be found. This work demonstrates, though, that the end result is well worth the effort, since it results in a physics-based and accurate compact model that is capable of predicting circuit performance and thus aiding in the deployment of a given technology.

Bibliography

[Ada82]   S. Adachi. Material parameters of In1- xGaxAsyP1- y and related binaries. *Journal of Applied Physics*, 53(12):8775–8792, 1982.

[Agi07]   Agilent Technologies. *Agilent PNA Microwave Network Analyzers Application Note 1408-12 - Pulsed-RF S-Parameter Measurement Using Wideband and Narrowband Detection*, 2007.

[Agi16]   Agilent Technologies. ADS 2016.01 online documentation, http://http://edadocs.software.keysight.com/pages/ viewpage.action?pageId=58332140. Online, 2016.

[Aur09]   Auriga Measurement Systems, LLC. *AU4750 Pulsed IV - Rulsed RF Measurement System User Guide*, 2009.

[BCS⁺02]   D. Berger, D. Céli, M. Schroter, M. Malorny, T. Zimmer, and B. Ardouin. HICUM parameter extraction methodology for a single transistor geometry. In *Proc. IEEE BCTM*, pages 116–119, 2002.

[BDE69]   W. E. Beadle, K. E. Daburlos, and W. H. Eckton. Design, fabrication, and characterization of a germanium microwave transistor. *IEEE Transactions on Electron Devices*, 16(1):125–138, Jan 1969.

[BFA⁺16]   C. R. Bolognesi, R. Fluckiger, M. Alexandrova, R. Lövblom, and O. Ostinelli. InP/GaInAsSb DHBT evolution in the THz era. In *2016 Compound Semiconductor Week (CSW)*, pages 1–4, June 2016.

[CSHB01]   S. V. Cherepko, M. S. Shirokov, J. C. M. Hwang, and A. Brandstaedter. Improved large-signal model and model extraction procedure for InGaP/GaAs HBTs under high-current operations.

In *IEEE International Microwave Symposium Digest*, volume 2, pages 671–674. IEEE, 2001.

[DB03]   M. W. Dvorak and C. R. Bolognesi. On the accuracy of direct extraction of the heterojunction-bipolar-transistor equivalent-circuit model parameters C_π, C_{BC}, and R_E. *IEEE Transactions on Microwave Theory and Techniques*, 51(6):1640–1649, 2003.

[DL92]   Paul Dodd and Mark Lundstrom. Minority electron transport in inp/ingaas heterojunction bipolar transistors. *Applied Physics Letters*, 61(4):465–467, 1992.

[DRM+11]   R. Driad, J. Rosenzweig, R. E. Makon, R. Losch, V. Hurm, H. Walcher, and M. Schlechtweg. InP DHBT-Based IC Technology for 100-Gb/s Ethernet. *IEEE Transactions on Electron Devices*, 58(8):2604–2609, Aug 2011.

[FHB+11]   A. Fox, B. Heinemann, R. Barth, S. Marschmeyer, C. Wipf, and Y. Yamamoto. SiGe: C HBT architecture with epitaxial external base. In *IEEE Bipolar/BiCMOS Circuits and Technology Meeting*, pages 70–73. IEEE, 2011.

[GCS]   GCS Inc. http://www.gcsincorp.com/. Online.

[GDR+05]   Z. Griffith, M. Dahlstrom, M. J. W. Rodwell, X. M. Fang, D. Lubyshev, Y. Wu, J. M. Fastenau, and W. K. Liu. InGaAs-InP DHBTs for increased digital IC bandwidth having a 391-GHz ft and 505-GHz fmax. *IEEE Electron Device Letters*, 26(1):11–13, Jan 2005.

[Gia72]   L. J. Giacoletto. Measurement of emitter and collector series resistances. *IEEE Transactions on Electron Devices*, 19(5):692–693, 1972.

[Gil10]   G. Gildenblat, editor. *Compact Modeling, Principles, Techniques and Applications*, chapter 7, "MEXTRAM". Springer, 2010. R. van der Toorn and J. C. J. Paasschens and W. J. Kloosterman and H. C. de Graaff.

[GP70]   H. K. Gummel and H. C. Poon. An integral charge control model of bipolar transistors. *Bell System Technical Journal*, 49(5):827–852, 1970.

[GR98]   R. Gabl and M. Reisch. Emitter series resistance from open-collector measurements-influence of the collector region and the parasitic pnp transistor. *IEEE Transactions on Electron Devices*, 45(12):2457–2465, 1998.

[GTB97]   Y. Gobert, P. J. Tasker, and K. H. Bachem. A physical, yet simple, small-signal equivalent circuit for the heterojunction bipolar transistor. *IEEE Transactions on Microwave Theory and Techniques*, 45(1):149–153, Jan 1997.

[GURP15]   Z. Griffith, M. Urteaga, P. Rowell, and R. Pierson. 340-440mW Broadband, High-Efficiency E-Band PA's in InP HBT. In *IEEE Compound Semiconductor Integrated Circuit Symposium (CSICS)*, pages 1–4, Oct 2015.

[HCN$^+$14]   M. Haferlach, M. Claus, T. Nardmann, A. Pacheco-Sanchez, P. Sakalas, and M. Schröter. Trap-induced apparent linearity of CNTFETs. In *Nanotech, Workshop on compact modeling*, 2014.

[HCZA96]   G. Hanington, C. E. Chang, P. J. Zampardi, and P. M. Asbeck. Thermal effects in HBT emitter resistance extraction. *Electronics Letters*, 32(16):1515–1516, 1996.

[HRB$^+$16]   B. Heinemann, H. Rücker, R. Barth, F. Bärwolf, J. Drews, G. G. Fischer, A. Fox, O. Fursenko, T. Grabolla, F. Herzel, J. Katzer, J. Korn, A. KrÃ$\frac{1}{4}$ger, P. Kulse, T. Lenke, M. Lisker, S. Marschmeyer, A. Scheit, D. Schmidt, J. Schmidt, M. A. Schubert, A. Trusch, C. Wipf, and D. Wolansky. SiGe HBT with ft/fmax of 505 GHz/720 GHz. In *2016 IEEE International Electron Devices Meeting (IEDM)*, pages 3.1.1–3.1.4, Dec 2016.

[HS09]   Z. Huszka and E. Seebacher. Extraction of RE and its temperature dependence from RF measurements. In *Working Group Bipolar (AKB) Meeting*, 2009.

[HSR$^+$12]   J. Herricht, P. Sakalas, M. Ramonas, M. Schroter, C. Jungemann, A. Mukherjee, and K. Moebus. Systematic compact modeling of correlated noise in bipolar transistors. *IEEE Transactions on Microwave Theory and Techniques*, 60(11):3403–3412, 2012.

[Hus05]   Z. Huszka. A Self Consistent Joint Extraction of τ_0 and RCX for HICUM. In *Proc. 6th HICUM Workshop*, 2005.

[HVvN04]   A. Huerta, T. Vanhoucke, and W. D. van Noort. Electrical characterization of high-performance SiGe: C HBTs focus on emitter-base junction. *Philips Res. Eindhoven, Eindhoven, The Netherlands, Tech. Rep. PR-TN-2004/00489*, 2004.

[HZC+07]   J. Hu, P. J. Zampardi, C. Cismaru, K. Kwok, and Y. Yang. Physics-Based Scalable Modeling of GaAs HBTs. In *IEEE Bipolar/BiCMOS Circuits and Technology Meeting*, pages 176–179, Sept 2007.

[IRS+03]   M. Iwamoto, D. E. Root, J. B. Scott, A. Cognata, P. M. Asbeck, B. Hughes, and D. C. D'Avanzo. Large-signal HBT model with improved collector transit time formulation for GaAs and InP technologies. In *IEEE International Microwave Symposium Digest*, volume 2, pages 635–638, June 2003.

[JDSC10]   J. Jacob, A. DasGupta, M. Schroter, and A. Chakravorty. Modeling nonquasi-static effects in SiGe HBTs. *IEEE Transactions on Electron Devices*, 57(7):1559–1566, 2010.

[JR91]   S. C. Jain and D. J. Roulston. A simple expression for band gap narrowing (BGN) in heavily doped Si, Ge, GaAs and $Ge_x Si_{1-x}$ strained layers. *Solid-State Electronics*, 34(5):453–465, 1991.

[Kaz13]   T. E. Kazior. More than Moore: III-V devices and Si CMOS get it together. In *IEEE International Electron Devices Meeting*, pages 28.5.1–28.5.4, Dec 2013.

[KML+08]   T. Kraemer, C. Meliani, F. Lenk, J. Würfl, and G. Tränkle. Transferred substrate DHBT of ft= 410 GHz and fmax= 480 GHz for traveling wave amplifiers. In *20th International Conference on Indium Phosphide and Related Materials*, 2008.

[KPK99]   W. J. Kloosterman, J. C. J. Paasschens, and D. B. M. Klaassen. Improved extraction of base and emitter resistance from small signal high frequency admittance measurements. In *Bipolar/BiCMOS Circuits and Technology Meeting, 1999. Proceedings of the 1999*, pages 93–96, 1999.

[Kra15]   J. Krause. *Model parameter extraction for very advanced heterojunction bipolar transistors*. PhD thesis, Chair for Electron Devices and Integrated Circuits, TU Dresden, 2015.

[Kro57]   H. Kroemer. Theory of a wide-gap emitter for transistors. *Proceedings of the IRE*, 45(11):1535–1537, 1957.

[KS15]   J. Krause and M. Schröter. Methods for Determining the Emitter Resistance in SiGe HBTs: A Review and an Evaluation Across Technology Generations. *IEEE Transactions on Electron Devices*, 62(5):1363–1374, May 2015.

[KU75]   K. F. Knott and R. T. Unwin. Comparison of noise-measured and theoretical value of base spreading resistance for an interdigitated transistor. *Electronics Letters*, 11(7):147–148, April 1975.

[LDB+94]   P. Launay, R. Driad, J. L. Benchimol, F. Alexandre, and J. Dangla. GaInP-GaAs Quasi Self-Aligned HBT Technology. In *ESSDERC'94: 24th European Solid State Device Research Conference*, pages 439–442. IEEE, 1994.

[LGG+98]   P. Llinares, G. Ghibaudo, N. Gambetta, Y. Mourier, A. Monroy, G. Lecoy, and J. A. Chroboczek. A novel method for base and emitter resistance extraction in bipolar junction transistors from static and low frequency noise measurements. In *Proceedings of the 1998 International Conference on Microelectronic Test Structures*, pages 67–71, Mar 1998.

[Log71]   J. Logan. Statistical Circuit Design: Characterization and Modeling for Statistical Design. *Bell System Technical Journal*, 50(4):1105–1147, 1971.

[Log72]   J. Logan. Modeling for circuit and system design. *Proceedings of the IEEE*, 60(1):78–85, Jan 1972.

[LWZ+11]   S. Lehmann, M. Weiss, Y. Zimmermann, A. Pawlak, K. Aufinger, and M. Schroter. Scalable compact modeling for SiGe HBTs suitable for microwave radar applications. In *IEEE 11th Topical Meeting on Silicon Monolithic Integrated Circuits in RF Systems (SiRF)*, pages 113–116. IEEE, 2011.

[McA03]   C. C. McAndrew. Status and Development of VBIC and Update on Verilog-A and ADMS. Technical report, TU Delft Workshop on Compact Modeling for RF/Microwave Applications, 2003.

[McA06]   C. C. McAndrew. BJT Base and Emitter Resistance Extraction from DC Data. In *Bipolar/BiCMOS Circuits and Technology Meeting*, pages 1–4, Oct 2006.

[MdRZ13]   A. G Metzger, V. dAlessandro, N. Rinaldi, and P. J. Zampardi. Evaluation of thermal balancing techniques in InGaP/GaAs HBT power arrays for wireless handset power amplifiers. *Microelectronics Reliability*, 53(9):1471–1475, 2013.

[MH82]   W. D. Mack and M. Horowitz. Measurement of series collector resistance in bipolar transistors. *IEEE Journal of Solid-State Circuits*, 17(4):767–773, 1982.

[MNT92]   S. A. Maas, B. L. Nelson, and D. L. Tait. Intermodulation in heterojunction bipolar transistors. *IEEE Transactions on Microwave Theory and Techniques*, 40(3):442–448, 1992.

[MSB+96]   C. C. McAndrew, J. A. Seitchik, D. F. Bowers, M. Dunn, M. Foisy, I. Getreu, M. McSwain, S. Moinian, J. Parker, D. J. Roulston, et al. VBIC95, the vertical bipolar inter-company model. *IEEE Journal of Solid-State Circuits*, 31(10):1476–1483, 1996.

[MT92]   S. A. Maas and D. Tait. Parameter-extraction method for heterojunction bipolar transistors. *IEEE Microwave and Guided Wave Letters*, 2(12):502–504, Dec 1992.

[Nar12]   T. Nardmann. Material models and parameters for InP, InGaAs and InGaAsP. Technical report, TU Dresden, CEDIC, 2012.

[Neu87]   A. Neugroschel. Measurement of the low-current base and emitter resistances of bipolar transistors. *IEEE Transactions on Electron Devices*, 34(4):817–822, Apr 1987.

[NH91]   T. Nakadai and K. Hashimoto. Measuring the base resistance of bipolar transistors. In *Proceedings of the 1991 Bipolar Circuits and Technology Meeting*, pages 200–203, Sep 1991.

[NKPS16]   T. Nardmann, J. Krause, A. Pawlak, and M. Schroter. Determining the base resistance of InP HBTs: An evaluation of methods and structures. *Solid-State Electronics*, 123:68–77, 2016.

[NKS14]   T. Nardmann, J. Krause, and M. Schroter. An Evaluation of Extraction Methods for the Emitter Resistance for InP DHBTs. In *IEEE Compound Semiconductor Integrated Circuit Symposium (CSICS)*, pages 1–4, Oct 2014.

[NLSD11]   T. Nardmann, S. Lehmann, M. Schröter, and R. Driad. Application of HICUM/L0 to InP DHBTs using single-transistor parameter extraction. In *23rd International Conference on Indium Phosphide and Related Materials (IPRM)*, pages 1–4. IEEE, 2011.

[NMPG09]   S. Nedeljkovic, J. McMacken, P. Partyka, and J. Gering. A Custom III-V Heterojunction Bipolar Transistor Model-Discussion of the features of a custom III-V heterojunction bipolar transistor model and its scaling for cellular power amplifier applications. *Microwave Journal; International ed*, 52(4):60, 2009.

[NS15]   T. Nardmann and M. Schroter. An evaluation of extraction methods for the external collector resistance for InP DHBTs. In *IEEE Compound Semiconductor Integrated Circuit Symposium (CSICS)*, pages 1–4. IEEE, 2015.

[NSC⁺13]   T. Nardmann, P. Sakalas, F. Chen, T. Rosenbaum, and M. Schroter. A Geometry Scalable Approach to InP HBT Compact Modeling for mm-Wave Applications. In *IEEE Compound Semiconductor Integrated Circuit Symposium (CSICS)*, pages 1–4, Oct 2013.

[NSSL13]   T. Nardmann, M. Schröter, P. Sakalas, and B. Lee. A length-scalable compact model for InP DHBTs. In *International Semiconductor Conference Dresden-Grenoble (ISCDG)*, pages 1–4, Sept 2013.

[NT84]   T. H. Ning and D. D. Tang. Method for determining the emitter and base series resistances of bipolar transistors. *IEEE Transactions on Electron Devices*, 31(4):409–412, Apr 1984.

[Paw14]   A. Pawlak. *Advanced Modeling of Silicon-Germanium Heterojunction Bipolar Transistors*. PhD thesis, Chair for Electron Devices and Integrated Circuits, TU Dresden, 2014.

[PLS14]   A. Pawlak, S. Lehmann, and M. Schroter. A simple and accurate method for extracting the emitter and thermal resistance of BJTs and HBTs. In *IEEE Bipolar/BiCMOS Circuits and Technology Meeting (BCTM)*, 2014.

[PQ04]   Vassil Palankovski and Rüdiger Quay. *Analysis and simulation of heterostructure devices*. Springer Science & Business Media, 2004.

[Pra92]   S. J. Prasad. A method of measuring base and emitter resistances of AlGaAs/GaAs HBTs. In *Proceedings of the 1992 Bipolar/BiCMOS Circuits and Technology Meeting*, pages 204–207, Oct 1992.

[PSS⁺12]   C. H. J. Poh, S. Seth, R. L. Schmid, J. D. Cressler, and J. Papapolymerou. Low-power and low-voltage x-band silicongermanium heterojunction bipolar transistor low-noise amplifier. *IET Microwaves, Antennas Propagation*, 6(12):1325–1331, September 2012.

[RAH11]   C. Raya, B. Ardouin, and Z. Huszka. Improving parasitic emitter resistance determination methods for advanced SiGe: C HBT transistors. In *IEEE Bipolar/BiCMOS Circuits and Technology Meeting*, pages 191–194. IEEE, 2011.

[RCC⁺15]   J. C. Rode, H. W. Chiang, P. Choudhary, V. Jain, B. J. Thibeault, W. J. Mitchell, M. J. W. Rodwell, M. Urteaga, D. Loubychev, A. Snyder, Y. Wu, J. M. Fastenau, and A. W. K. Liu. Indium Phosphide Heterobipolar Transistor Technology Beyond 1-THz Bandwidth. *IEEE Transactions on Electron Devices*, 62(9):2779–2785, Sept 2015.

[RD15]   W. Y. Refai and W. A. Davis. A linear, highly-efficient, class-J handset power amplifier utilizing GaAs HBT technology. In *IEEE 16th Annual Wireless and Microwave Technology Conference (WAMICON)*, pages 1–4, April 2015.

[Rei84]   H.-M. Rein. A simple method for separation of the internal and external (peripheral) currents of bipolar transistors. *Solid-State Electronics*, 27(7):625–631, 1984.

[RHB⁺08]   D. E. Root, J. Horn, L. Betts, C. Gillease, and J. Verspecht. X-parameters: The new paradigm for measurement, modeling, and design of nonlinear RF and microwave components. *Microwave Engineering Europe*, pages 16–21, 2008.

[RHF12]   H. Rücker, B. Heinemann, and A. Fox. Half-Terahertz SiGe BiCMOS technology. In *2012 IEEE 12th Topical Meeting on Silicon Monolithic Integrated Circuits in RF Systems*, pages 133–136, Jan 2012.

[RHFS94]   D. Ritter, R.A. Hamm, A. Feygenson, and P.R. Smith. Role of hot electron base transport in abrupt emitter InP/Ga0. 43In0. 53As heterojunction bipolar transistors. *Applied physics letters*, 64(22):2988–2990, 1994.

[RKP⁺07]   C. Raya, N. Kauffmann, F. Pourchon, D. Celi, and T. Zimmer. Scalable approach for external collector resistance calculation. In *IEEE International Conference on Microelectronic Test Structures*, pages 101–106. IEEE, 2007.

[RLB08]   M. J. W. Rodwell, M. Le, and B. Brar. InP Bipolar ICs: Scaling Roadmaps, Frequency Limits, Manufacturable Technologies. *Proceedings of the IEEE*, 96(2):271–286, Feb 2008.

[Roh02]   M. Rohner. *Physical limitations of InP/InGaAs heterojunction-bipolar transistors*. PhD thesis, Diss., Technische Wissenschaften ETH Zürich, Nr. 14508, 2002.

[Ros16]   T. Rosenbaum. *Performance prediction of a future SiGe HBT technology using a heterogeneous set of simulation tools and approaches*. PhD thesis, Technische Universität Dresden, Université de Bordeaux, 2016.

[RPHJ06]   J. M. Ruiz-Palmero, U. Hammer, and H. Jäckel. A physical hydrodynamic 2D model for simulation and scaling of InP/InGaAs (P) DHBTs and circuits with limited complexity. *Solid-State Electronics*, 50(9):1595–1611, 2006.

[RPHJ$^+$07]   J. M. Ruiz-Palmero, U. Hammer, H. Jäckel, H. Liu, and C. R. Bolognesi. Comparative technology assessment of future InP HBT ultrahigh-speed digital circuits. *Solid-State Electronics*, 51(6):842–859, 2007.

[RRG$^+$12]   T. B. Reed, M. Rodwell, Z. Griffith, P. Rowell, A. Young, M. Urteaga, and M. Field. A 220 GHz InP HBT Solid-State Power Amplifier MMIC with 90mW POUT at 8.2 dB Compressed Gain. In *IEEE Compound Semiconductor Integrated Circuit Symposium (CSICS)*, pages 1–4. IEEE, 2012.

[RS91]   H.-M. Rein and M. Schröter. Experimental determination of the internal base sheet resistance of bipolar transistors under forward-bias conditions. *Solid-State Electronics*, 34(3):301 – 308, 1991.

[RSD$^+$11]   A. Rumiantsev, P. Sakalas, N. Derrier, D. Celi, and M. Schroter. Influence of probe tip calibration on measurement accuracy of small-signal parameters of advanced BiCMOS HBTs. In *IEEE Bipolar/BiCMOS Circuits and Technology Meeting*, pages 203–206. IEEE, 2011.

[RSPL13]   T. Rosenbaum, M. Schröter, A. Pawlak, and S. Lehmann. Automated transit time and transfer current extraction for single transistor geometries. In *2013 IEEE Bipolar/BiCMOS Circuits and Technology Meeting (BCTM)*, pages 25–28. IEEE, 2013.

[Rud05]   M. Rudolph. FBH HBT model, hppt://www.fbh-berlin.de/modeling.html. Online, 2005.

[Rud06]   M. Rudolph. *Introducton to modeling HBTs*. Artech House, 2006.

[Sak16]   P. Sakalas. Private communications, 2016.

[SC10]   M. Schroter and A. Chakravorty. *Compact hierarchical modeling of bipolar transistors with HICUM*. World Scientific, Singapore, 2010.

[Sch88]   M. Schröter. *A compact physical large-signal model for high-speed bipolar transistors with special regard to high current densities and two-dimensional effects*. PhD thesis, Ruhr-University Bochum, 1988.

[Sch90]   D. Schroeder. Three-dimensional nonequilibrium interface conditions for electron transport at band edge discontinuities. *IEEE Transactions on Computer-Aided Design of Integrated Circuits and Systems*, 9(11):1136–1140, 1990.

[Sch91a]   M. Schroter. Simulation and modeling of the low-frequency base resistance of bipolar transistors and its dependence on current and geometry. *IEEE Transactions on Electron Devices*, 38(3):538–544, Mar 1991.

[Sch91b]   M. Schröter. Transient and small-signal high-frequency simulation of numerical device models embedded in an external circuit. *COMPEL-The international journal for computation and mathematics in electrical and electronic engineering*, 10(4):377–387, 1991.

[Sch08]   M. Schröter. *DEVICE - A Mixed-Mode Simulator for Three-Dimensional Heterostructure Semiconductor Devices and Circuits*. TU Dresden, CEDIC, 2.03 edition, 2008.

[Sch13]   M. Schroter. Geometry scaling. Technical report, TU Dresden, CEDIC, 2013.

[SCS$^+$13]   M. Schroter, M. Claus, P. Sakalas, M. Haferlach, and D. Wang. Carbon nanotube FET technology for radio-frequency electronics: state-of-the-art overview. *IEEE J. Electron Devices Soc*, 1(1):9–20, 2013.

[SKLC08]   M. Schroter, J. Krause, S. Lehmann, and D. Celi. Compact Layout and Bias-Dependent Base-Resistance Modeling for Advanced SiGe HBTs. *IEEE Transactions on Electron Devices*, 55(7):1693–1701, July 2008.

[SKR00]   M. Sotoodeh, A.H. Khalid, and A.A. Rezazadeh. Empirical low-field mobility model for III–V compounds applicable in device simulation codes. *Journal of Applied Physics*, 87(6):2890–2900, 2000.

[SKW+11]   M. Schroter, P. Kolev, D. Wang, M. Eron, S. Lin, N. Samarakone, M. Bronikowski, Z. Yu, P. Sampat, P. Syams, and S. McKernan. A 4" wafer photostepper-based carbon nanotube FET technology for RF applications. In *IEEE International Microwave Symposium Digest*, pages 1–4, June 2011.

[SL07]   M. Schroter and S. Lehmann. The rectangular bipolar transistor tetrode structure and its application. In *IEEE International Conference on Microelectronic Test Structures*, pages 206–209, March 2007.

[SM72]   W. M. C. Sansen and R. G. Meyer. Characterization and measurement of the base and emitter resistances of bipolar transistors. *IEEE Journal of Solid-State Circuits*, 7(6):492–498, Dec 1972.

[SNS+16]   P. Sakalas, T. Nardmann, A. Simukovic, M. Schröter, and H. Zirath. Microwave noise analysis in inp and gaas hbts (accepted for publication). In *Proc. IEEE CSICS*, 2016.

[SNZ]   M. Schröter, T. Nardmann, and P.J. Zampardi. A closed-form solution for the low-current collector transit time in group IV and group III-V HBTs. (in progress).

[SP14]   M. Schroter and A. Pawlak. Physics-based nonlinear compact modeling of HBTs for mm-wave applications. In *2014 International Microwave Symposium*, Tampa Bay, Florida, USA, 2014.

[SPM15]   M. Schröter, A. Pawlak, and A. Mukherjee. HICUM/L2 - A geometry scalable physics-based compact bipolar transistor model - documentation of model version 2.34. Technical report, TU Dresden, CEDIC, 2015.

[SRC+16]   M. Schröter, T. Rosenbaum, P. Chevalier, B. Heinemann, S. P. Voinigescu, E. Preisler, J. Böck, and A. Mukherjee. SiGe HBT Technology: Future Trends and TCAD-Based Roadmap. *Proceedings of the IEEE*, PP(99):1–19, 2016.

[SRR+99]   M. Schroter, H.-M. Rein, W. Rabe, R. Reimann, H.-J. Wassener, and A. Koldehoff. Physics-and process-based bipolar transistor modeling for integrated circuit design. *IEEE Journal of Solid-State Circuits*, 34(8):1136–1149, 1999.

[SS13]   P. Sakalas and M. Schroter. Microwave noise in InP and SiGe HBTs: Modeling and challenges. In *22nd International Conference on Noise and Fluctuations (ICNF)*, pages 1–6. IEEE, 2013.

[SSP80]   S. Selberherr, A. Schutz, and H. W. Potzl. MINIMOS-A two-dimensional MOS transistor analyzer. *IEEE Journal of Solid-State Circuits*, 15(4):605–615, 1980.

[ST06]   M. Schroter and H. Tran. Charge-storage related parameter calculation for Si and SiGe bipolar transistors from device simulation. In *Proc. WCM, International NanoTech Meeting*, pages 735–740, 2006.

[SUH⁺11]   M. Seo, M. Urteaga, J. Hacker, A. Young, Z. Griffith, V. Jain, R. Pierson, P. Rowell, A. Skalare, A. Peralta, et al. InP HBT IC technology for terahertz frequencies: Fundamental oscillators up to 0.57 THz. *IEEE Journal of Solid-State Circuits*, 46(10):2203–2214, 2011.

[SUZC15]   R. L. Schmid, A. C. Ulusoy, S. Zeinolabedinzadeh, and J. D. Cressler. A Comparison of the Degradation in RF Performance Due to Device Interconnects in Advanced SiGe HBT and CMOS Technologies. *IEEE Transactions on Electron Devices*, 62(6):1803–1810, June 2015.

[SWH⁺11]   M. Schroter, G. Wedel, B. Heinemann, C. Jungemann, J. Krause, P. Chevalier, and A. Chantre. Physical and Electrical Performance Limits of High-Speed SiGeC HBTs Part I: Vertical Scaling. *IEEE Transactions on Electron Devices*, 58(11):3687–3696, Nov 2011.

[Syn15]   Synopsis. *Sentaurus Device User Guide Version K-2015.06*, 2015.

[Tik16]   P. Tikhomirov. Private communication, 2016.

[TLB05]   N. G. M. Tao, H. G. Liu, and C. R. Bolognesi. Surface recombination currents in" Type-II" NpN InP-GaAsSb-InP self-aligned DHBTs. *IEEE Transactions on Electron Devices*, 52(6):1061–1066, 2005.

[TLB07]   N. G. M. Tao, H. G. Liu, and C. R. Bolognesi. Two-dimensional simulation of type-II InP/GaAsSb/InP double heterojunction bipolar transistors. *Physica Status Solidi (c)*, 4(5):1675–1679, 2007.

[TSW+97] H. Tran, M. Schroter, D. J. Walkey, D. Marchesan, and T. J. Smy. Simultaneous extraction of thermal and emitter series resistances in bipolar transistors. In *Bipolar/BiCMOS Circuits and Technology Meeting*, pages 170–173. IEEE, 1997.

[UCS00] UCSD. HBT Model Equations rev 9.001a 3/2000, http://hbt.ucsd.edu/. Online, 2000.

[UPR+11] M. Urteaga, R. Pierson, P. Rowell, V. Jain, E. Lobisser, and M. J. W. Rodwell. 130nm InP DHBTs with ft¿ 0.52 THz and fmax¿ 1.1 THz. In *69th Annual Device Research Conference (DRC)*, pages 281–282. IEEE, 2011.

[Urt16] M. Urteaga. Private communications, 2016.

[VCT+93] G. Verzellesi, A. Chantre, R. Turetta, M. Cappellin, P. Pavan, and E. Zanoni. A Compact Method for Measuring Parasitic Resistances in Bipolar Transistors. In *ESSDERC '93: 23rd European solid State Device Research Conference*, pages 433–436, Sept 1993.

[VDK+96] V. Van, M. J. Deen, J. Kendall, D. S. Malhi, S. Voinigescu, and M. Schrofer. DC extraction of the base and emitter resistances in polysilicon-emitter npn BJTs. *Canadian Journal of Physics*, 74(S1):172–176, 1996.

[VTD+13] S.P. Voinigescu, A. Tomkins, E. Dacquay, P. Chevalier, J. Hasch, A. Chantre, and B. Sautreuil. A study of SiGe HBT signal sources in the 220–330-GHz range. *IEEE Journal of Solid-State Circuits*, 48(9):2011–2021, 2013.

[WCS+14] D. F. Williams, P. Corson, J. Sharma, H. Krishnaswamy, W. Tai, Z. George, D. S. Ricketts, P. M. Watson, E. Dacquay, and S. P. Voinigescu. Calibrations for millimeter-wave silicon transistor characterization. *IEEE Transactions on Microwave Theory and Techniques*, 62(3):658–668, 2014.

[Wed16a] G. Wedel. *A Box-Integration/WENO solver for the Boltzmann Transport Equation and its Application to High-Speed HBTs*. PhD thesis, TU Dresden, Chair for electron devices and integrated circuits, 2016.

[Wed16b]   G. Wedel. Terahertz sige hbt tcad and numerical device simulation. In *Short course "Terahertz SiGe HBT technologies: Creating the winning design engine" at IEEE CSICS*, 2016.

[WMX$^+$06]   N. L. Wang, W. Ma, S. Xu, E. Camargo, X. Sun, P. Hu, Z. Tang, H. F. F. Chau, A. Chen, and C. P. Lee. 28V High-Linearity and Rugged InGaP/GaAs Power HBT. In *IEEE MTT-S International Microwave Symposium Digest*, pages 881–884, June 2006.

[Yu70]   A. Y. C. Yu. Electron tunneling and contact resistance of metal-silicon contact barriers. *Solid-State Electronics*, 13(2):239–247, 1970.

[YZ13]   Y. Yang and P. J. Zampardi. Base Resistance Scaling for Transistors of Various Geometries. In *IEEE Compound Semiconductor Integrated Circuit Symposium (CSICS)*, pages 1–4, Oct 2013.

[YZK15]   Y. Yang, P. J. Zampardi, and K. Kwok. HBT Model Scaling for Different Epi Materials and Geometries. In *IEEE Compound Semiconductor Integrated Circuit Symposium (CSICS)*, pages 1–4. IEEE, 2015.

[Zam16]   P. Zampardi. Private communications, 2016.

[ZBB$^+$96]   T. Zimmer, J. Berkner, B. Branciard, N. Lewis, J. B. Duluc, and J. P. Dom. Method for determining the effective base resistance of bipolar transistors. In *Bipolar/BiCMOS Circuits and Technology Meeting*, pages 122–125, Sep 1996.

[ZCS$^+$04]   G. Zohar, S. Cohen, V. Sidorov, A. Gavrilov, B. Sheinman, and D. Ritter. Reduction of base-transit time of InP-GaInAs HBTs due to electron injection from an energy ramp and base-composition grading. *IEEE Transactions on Electron Devices*, 51(5):658–662, 2004.

[ZHY$^+$07]   P. Zampardi, J. Hu, Y. Yang, K. Kwok, C. Cismaru, and M. Banbrook. Industrial Practices of Characterization and Modeling of III-V Technology. In *Workshop on Compact Modeling for RF/Microwave Applications*, 2007.

[Zim16]   Y. Zimmermann. *Contributions to the analytical modeling of distributed thermal and electrical substrate coupling effects in HBTs*. PhD thesis, TU Dresden, Chair for electron devices and integrated circuits, 2016.

[Zir12] H. Zirath. Private communications, 2012.

[ZKC$^+$12] P. J. Zampardi, K. Kwok, C. Cismaru, M. Sun, and A. Lo. Characteristics of GaAs spike doped collectors. In *IEEE Bipolar/BiCMOS Circuits and Technology Meeting (BCTM)*, pages 1–4, Sept 2012.

[ZMCD91] T. Zimmer, A. Meresse, Ph. Cazenave, and J.P. Dom. Simple determination of BJT extrinsic base resistance. *Electronics Letters*, 27(21):1895 – 1896, 1991.

[ZMWS09] Y. Zimmermann, K. Moebus, H. Wittkopf, and M. Schroter. TRADICA - An integrated modeling tool linking process and circuit design. In *IEEE SCD*, page 4, 2009.

APPENDIX A

Material parameters for numerical device simulations

This section lists the material parameters used in the numerical in-house simulator CHIEF, obtained by fitting the listed material models to experimental data where available, Monte-Carlo data where not and utilizing parameters directly from literature otherwise; using the interpolation scheme described in section 2.4. The full list of references for the data used to determine the parameters can be found in [Nar12].

A.1 Band gap

The band gap is modeled using the equation

$$E_{\mathrm{g}} = E_{\mathrm{g0}} - \frac{A \cdot T^2}{T + B},$$
(A.1.0-1)

with E_{g0}, the bandgap at $T = 0\,\mathrm{K}$, and the model parameters A and B.

Table A.1: Composition dependence of the parameters for the band gap

	$\mathrm{In_{1\text{-}x}Ga_xAs}$			$\mathrm{In_{1\text{-}x}Ga_xAs_yP_{1\text{-}y}},\ x = 0.47y$		
	a	b	c	a	b	c
$E_{\mathrm{g0}}/\mathrm{eV}$	0.415	0.654	0.45	1.421	-0.9344	0.3352
$A/10^4$	2.76	11.34	-8.70	5.36	9.04	-8.24
B/K	83	551.96	-430.96	327	85.96	-165.73

Table A.2: Composition dependence of the parameters for the band gap

	$In_{1-x}Ga_xAs$			$In_{1-x}Ga_xAs_yP_{1-y}, x = 0.47y$		
	a	b	c	a	b	c
$\Delta\chi_0/eV$	0.599	0.240	0.165	0.217	-0.363	0.130
$A/10^4$	1.01	3.81	-2.84	1.90	4.15	-3.75
B/K	83	478.5	-357.6	327	102.7	-200.8

Table A.3: Fractional variation of the conduction resp. valence band with composition

	$In_{1-x}Ga_xAs$	$In_{1-x}Ga_xAs_yP_{1-y}, x = 0.47$
Conduction band	$0.367\ \Delta E_g$	$0.388\ \Delta E_g$
Valence band	$0.633\ \Delta E_g$	$0.612\ \Delta E_g$

The electron affinity, i.e. the difference in conduction band energies, is accordingly modeled as

$$\Delta\chi = \Delta\chi_0 - \frac{A \cdot T^2}{T + B}, \tag{A.1.0-2}$$

with the electron affinity at $T = 0\,\mathrm{K}$, $\Delta\chi_0$ and the model parameters A and B.

Alternatively, the band differences can be calculated based on the band gap change w.r.t. the base material (x=0) according to tab. A.3.

A.2   Band gap narrowing

The band gap narrowing is modeled according to [JR91] as

$$\Delta E_g = A \cdot N^{1/3} + B \cdot N^{1/4} + C \cdot N^{1/2}, \tag{A.2.0-3}$$

with the doping concentration N and the model parameters A, B and C. The parameters differ for n-type and p-type doping. For the material system GaAs-InAs, the interpolation from the available data points results in physically meaningless parameters (i.e. negative bandgap narrowing) for a certain composition range. Instead of allowing this, the interpolation is split in two intervals.

Table A.4: Band gap narrowing for the InP-InGaAs material system, n-type

	$\mathrm{In_{1-x}Ga_xAs_yP_{1-y}}, x = 0.47y$		
	a	b	c
$A/(\mathrm{eVcm})$	$6.826 \cdot 10^{-8}$	$6.826 \cdot 10^{-8}$	0
$B/(\mathrm{eVcm}^{3/4})$	0	0	0
$C/(\mathrm{eVcm}^{3/2})$	$1.6383 \cdot 10^{-12}$	$4.3407 \cdot 10^{-11}$	0

Table A.5: Band gap narrowing for the InP-InGaAs material system, p-type

	$\mathrm{In_{1-x}Ga_xAs_yP_{1-y}}, x = 0.47y$		
	a	b	c
$A/(\mathrm{eVcm})$	$1.03 \cdot 10^{-8}$	$-1.1 \cdot 10^{-9}$	0
$B/(\mathrm{eVcm}^{3/4})$	$4.43 \cdot 10^{-7}$	$-8.6 \cdot 10^{-8}$	0
$C/(\mathrm{eVcm}^{3/2})$	$3.38 \cdot 10^{-12}$	$2.70 \cdot 10^{-13}$	0

Table A.6: Band gap narrowing for the GaAs-InAs material system, n-type

	x	$\mathrm{In_{1-x}Ga_xAs}$		
		a	b	c
$A/(\mathrm{eVcm})$	$0\ldots0.47$	$1.4 \cdot 10^{-8}$	$-5.96 \cdot 10^{-8}$	$6.34 \cdot 10^{-8}$
$B/(\mathrm{eVcm}^{3/4})$	$0\ldots0.47$	$1.97 \cdot 10^{-7}$	$-8.38 \cdot 10^{-7}$	$8.92 \cdot 10^{-7}$
$C/(\mathrm{eVcm}^{3/2})$	$0\ldots0.47$	$5.79 \cdot 10^{-11}$	$-5.47 \cdot 10^{-11}$	$-5.819 \cdot 10^{-11}$
$A/(\mathrm{eVcm})$	$0.47\ldots1$	0	0	0
$B/(\mathrm{eVcm}^{3/4})$	$0.47\ldots1$	$2.00 \cdot 10^{-6}$	$-8.49 \cdot 10^{-6}$	$9.03 \cdot 10^{-6}$
$C/(\mathrm{eVcm}^{3/2})$	$0.47\ldots1$	$9.64 \cdot 10^{-12}$	$-1.51 \cdot 10^{-10}$	$-1.60 \cdot 10^{-10}$

Table A.7: Band gap narrowing for the GaAs-InAs material system, p-type

	$\mathrm{In_{1-x}Ga_xAs_yP_{1-y}}, x = 0.47y$		
	a	b	c
$A/(\mathrm{eVcm})$	$8.34 \cdot 10^{-9}$	$1.09 \cdot 10^{-8}$	$-1.92 \cdot 10^{-8}$
$B/(\mathrm{eVcm}^{3/4})$	$2.91 \cdot 10^{-7}$	$5.23 \cdot 10^{-7}$	$-8.14 \cdot 10^{-7}$
$C/(\mathrm{eVcm}^{3/2})$	$4.53 \cdot 10^{-12}$	$-5.16 \cdot 10^{-12}$	$7.00 \cdot 10^{-12}$

A.3 Effective mass

Table A.8: Composition dependence of the effective carrier mass

	$\mathrm{In}_{1\text{-}x}\mathrm{Ga}_x\mathrm{As}$			$\mathrm{In}_{1\text{-}x}\mathrm{Ga}_x\mathrm{As}_y\mathrm{P}_{1\text{-}y}, x = 0.47y$		
	a	b	c	a	b	c
$m_{\mathrm{m,eff}}/m_0$	0.023	0.037	0.003	0.08	-0.0390	0
$m_{\mathrm{m,eff}}/m_0$	0.41	0.1	0	0.6	-0.143	0

A.4 Dielectric constant

Table A.9: Composition dependence of the dielectric constant

	$\mathrm{In}_{1\text{-}x}\mathrm{Ga}_x\mathrm{As}$			$\mathrm{In}_{1\text{-}x}\mathrm{Ga}_x\mathrm{As}_y\mathrm{P}_{1\text{-}y}, x = 0.47y$		
	a	b	c	a	b	c
ε_r	15.1	-2.87	0.67	12.5	1.44	0

A.5 Impact ionization

The impact ionization rate in the simulator is described as [Sch08]

$$\alpha_\mathrm{c} = \alpha_{\mathrm{c,inf}} \cdot \exp\left(-\frac{F_{\mathrm{c,av}}}{F_\mathrm{c}}\right), \tag{A.5.0-4}$$

with the model parameters $F_{\mathrm{c,av}}$ and $\alpha_{\mathrm{c,inf}}$ and the gradient of the respective quasi-fermi potential F_c.

Table A.10: Electron impact ionization parameters for the InP-InGaAs material system

	$\mathrm{In}_{1\text{-}x}\mathrm{Ga}_x\mathrm{As}_y\mathrm{P}_{1\text{-}y}, x = 0.47y$		
	a	b	c
$\alpha_{\mathrm{c,inf}}/(\mathrm{cm})$	$3.886 \cdot 10^6$	$4.667 \cdot 10^6$	$-8.389 \cdot 10^6$
$F_{\mathrm{c,av}}/(\mathrm{Vcm})$	$2.78 \cdot 10^6$	$1.50 \cdot 10^6$	$-3.28 \cdot 10^6$

Table A.11: Hole impact ionization parameters for the InP-InGaAs material system

	$\text{In}_{1\text{-x}}\text{Ga}_{\text{x}}\text{As}_{\text{y}}\text{P}_{1\text{-y}}, x = 0.47y$		
	a	b	c
$\alpha_{c,inf}/(cm)$	$2.900 \cdot 10^6$	$1.357 \cdot 10^7$	$-1.631 \cdot 10^7$
$F_{c,av}/(Vcm)$	$2.40 \cdot 10^6$	$4.25 \cdot 10^6$	$-5.85 \cdot 10^6$

Table A.12: Electron impact ionization parameters for the GaAs-InAs material system

	x	$\text{In}_{1\text{-x}}\text{Ga}_{\text{x}}\text{As}$		
		a	b	c
$\alpha_{c,inf}/(cm)$	$0\ldots0.47$	$2.14 \cdot 10^5$	$-2.13 \cdot 10^5$	$2.27 \cdot 10^5$
$F_{c,av}/(Vcm)$	$0\ldots0.47$	$1.60 \cdot 10^6$	$-2.57 \cdot 10^6$	$2.73 \cdot 10^6$
$\alpha_{c,inf}/(cm)$	$0.47\ldots1$	$1.35 \cdot 10^8$	$-5.73 \cdot 10^8$	$6.09 \cdot 10^8$
$F_{c,av}/(Vcm)$	$0.47\ldots1$	$3.16 \cdot 10^6$	$-9.20 \cdot 10^6$	$9.79 \cdot 10^6$

Table A.13: Hole impact ionization parameters for the GaAs-InAs material system

	$\text{In}_{1\text{-x}}\text{Ga}_{\text{x}}\text{As}$		
	a	b	c
$\alpha_{c,inf}/(cm)$	$1.15 \cdot 10^6$	$-5.32 \cdot 10^6$	$6.83 \cdot 10^6$
$F_{c,av}/(Vcm)$	$1.05 \cdot 10^6$	$-1.94 \cdot 10^6$	$2.99 \cdot 10^6$

A.6 Recombination

The recombination rates for Auger and radiative recombination are only composition-dependent. The carrier lifetimes for SRH recombination are, in theory, both doping- and composition-dependent; however, insufficient data for extracting the doping dependence was available. Similarly, no separate parameters for electron and hole lifetimes could be found.

The recombination rates are calculates as

$$R_{\text{SRH}} = \frac{pn - n_i^2}{\tau_n(n_i + p) + \tau_p(n_i + n)}, \tag{A.6.0-5}$$

$$R_{\text{aug}} = R_{n,aug}n(np - n_i^2) + R_{p,aug}p(np - n_i^2) \text{and} \tag{A.6.0-6}$$

$$R_{\text{rad}} = R_{np,rad}(np - n_i^2) + R_{p,aug}p(np - n_i^2), \tag{A.6.0-7}$$

with the carrier densities n and p and the intrinsic carrier density n_i.

Table A.14: Auger recombination coefficients for the GaAs-InAs material system

	x	$\mathrm{In_{1-x}Ga_xAs}$		
		a	b	c
$R_{\mathrm{n,aug}}/(\mathrm{s/cm^6})$	$0\ldots0.47$	$6.6\cdot10^{-27}$	$-2.8\cdot10^{-26}$	$2.95\cdot10^{-26}$
$R_{\mathrm{p,aug}}/(\mathrm{s/cm^6})$	$0\ldots0.47$	$2.2\cdot10^{-27}$	$-9.0\cdot10^{-27}$	$9.59\cdot10^{-27}$
$R_{\mathrm{n,aug}}/(\mathrm{s/cm^6})$	$0.47\ldots1$	$2.3\cdot10^{-29}$	$2.5\cdot10^{-28}$	$-2.6\cdot10^{-28}$
$R_{\mathrm{p,aug}}/(\mathrm{s/cm^6})$	$0.47\ldots1$	$2.3\cdot10^{-29}$	$2.5\cdot10^{-28}$	$-2.6\cdot10^{-28}$

Table A.15: Auger recombination coefficients for the InP-InGaAs material system

	$\mathrm{In_{1-x}Ga_xAs_yP_{1-y}},\ x=0.47y$		
	a	b	c
$R_{\mathrm{n,aug}}/(\mathrm{s/cm^6})$	$1.6\cdot10^{-30}$	$1.7\cdot10^{-28}$	$-8.8\cdot10^{-29}$
$R_{\mathrm{p,aug}}/(\mathrm{s/cm^6})$	$9.0\cdot10^{-31}$	$1.7\cdot10^{-28}$	$9.0\cdot10^{-29}$

Table A.16: Radiative recombination coefficients

	$\mathrm{In_{1-x}Ga_xAs}$			$\mathrm{In_{1-x}Ga_xAs_yP_{1-y}},\ x=0.47y$		
	a	b	c	a	b	c
$R_{\mathrm{np,rad}}/(\mathrm{s/cm^3})$	$9.8\cdot10^{-11}$	0	0	$1.1\cdot10^{-10}$	$-1.2\cdot10^{-11}$	0

Table A.17: SRH carrier lifetimes

	$\mathrm{In_{1-x}Ga_xAs}$			$\mathrm{In_{1-x}Ga_xAs_yP_{1-y}},\ x=0.47y$		
	a	b	c	a	b	c
$\tau_{\mathrm{max}}/\mathrm{s}$	$4.7\cdot10^{-5}$	0	0	$4\cdot10^{-8}$	$4.696\cdot10^{-5}$	0

A.7 Low-field mobility

The doping- and temperature low-field mobility is calculated as [SKR00]

$$\mu(N,T) = \mu_{\min} + \frac{\mu_{\max}\left(\frac{300\,\mathrm{K}}{T}\right)^{\theta_1} - \mu_{\min}}{1 + \left(\dfrac{N}{N_{\mathrm{ref}}\left(\frac{T}{300\,\mathrm{K}}\right)^{\theta_2}}\right)^{\lambda}} \tag{A.7.0-8}$$

The mobility values $\mu_{\min}$ and $\mu_{\max}$ are interpolated from the parameters in the tables as their inverse according to Matthiesen's rule,

$$\mu = \left(\mu_{\mathrm{a}} + x\mu_{\mathrm{b}} + x^2\mu_{\mathrm{c}}\right)^{-1}. \tag{A.7.0-9}$$

Table A.18: Low-field electron mobility parameters

	$\mathrm{In_{1-x}Ga_xAs}$			$\mathrm{In_{1-x}Ga_xAs_yP_{1-y}}, x = 0.47y$		
	a	b	c	a	b	c
$\mu_{\mathrm{n,min}}/(10^4\,\mathrm{cm}^2/(Vs))$	5.08	-11	66	12.1	17.7	-15.2
$\mu_{\mathrm{n,max}}/(10^4\,\mathrm{cm}^2/(Vs))$	0.16	2.35	-1.21	1.93	2.85	-3.79
λ	0.48	1.75	-1.79	0.55	-1.58	1.94
$N_{\mathrm{ref}}/(10^{17}\,\mathrm{cm}^{-3})$	1	27.2	-21.7	2	18.7	-11.7
θ_1	3.11	-10.06	10.16	4.15	-7.07	3.56
θ_2	8	-27.14	26.94	8	-2.29	-4.51

Table A.19: Low-field hole mobility parameters

	$\mathrm{In_{1-x}Ga_xAs}$			$\mathrm{In_{1-x}Ga_xAs_yP_{1-y}}, x = 0.47y$		
	a	b	c	a	b	c
$\mu_{\mathrm{p,min}}/(10^2\,\mathrm{cm}^2/(Vs))$	1.41	-2.67	5.54	1.8	35.7	-36.1
$\mu_{\mathrm{p,max}}/(10^3\,\mathrm{cm}^2/(Vs))$	2.1	-0.22	0.32	5.54	39.4	-42.9
λ	0.58	2.61	-2.75	0.46	-0.18	0.92
$N_{\mathrm{ref}}/(10^{17}\,\mathrm{cm}^{-3})$	2	14.1	-14.1	1.1	-3.927	8.32
θ_1	3.65	-7.17	9.02	3.65	-4.73	1.64
θ_2	4.02	-18.72	21.95	5.26	1.95	-7.14

A.8 Field-dependent mobility

The field-dependent mobility is calculated according to [SSP80]

$$\mu(F_c) = \frac{\mu_G + v_{sat}\frac{F_c^{\beta-1}}{F_0^\beta}}{1 + \frac{F_c^\beta}{F_0^\beta}}. \tag{A.8.0-10}$$

The temperature-dependent saturation velocity is described by [Syn15]

$$v_{sat} = v_{sat,0} - v_{sat,T}\left(\frac{T}{T_0}\right). \tag{A.8.0-11}$$

Since the goal is primarily to simulate npn transistors and there is little data available in the literature regarding the hole mobility, the critical field for the holes is set to a constant value of $F_{0,p} = 100\,\text{kV}\,\text{cm}^{-1}$ and the exponential coefficient to $\beta_p = 1$.

Table A.20: High-field electron mobility parameters

	$\text{In}_{1\text{-x}}\text{Ga}_x\text{As}$			$\text{In}_{1\text{-x}}\text{Ga}_x\text{As}_y\text{P}_{1\text{-y}}, x = 0.47y$		
	a	b	c	a	b	c
β_n	2.065	5.318	-3.217	3.63	-6.49	6.71
$F_{0,n}/(\text{kV}/(\text{cm}))$	1.86	7.1	-5.02	14	-18.1	8.16
$v_{n,sat,0}/(10^7\,\text{cm/s})$	-0.224	5	-3.95	0.89	-1.37	1.75
$v_{n,sat,T}/(10^7\,\text{cm/s})$	0.198	-2.08	0.443	0.266	-0.601	0.533

Table A.21: High-field hole mobility parameters

	$\text{In}_{1\text{-x}}\text{Ga}_x\text{As}$			$\text{In}_{1\text{-x}}\text{Ga}_x\text{As}_y\text{P}_{1\text{-y}}, x = 0.47y$		
	a	b	c	a	b	c
β_n	1	0	0	1	0	0
$F_{0,n}/(\text{kV}/(\text{cm}))$	100	0	0	100	0	0
$v_{n,sat,0}/(10^5\,\text{cm/s})$	5	141	-46.8	70	-8.9	0
$v_{n,sat,T}/(10^5\,\text{cm/s})$	0	0	0	0	0	0

A.9 Energy relaxation time

The energy-dependent energy relaxation time follows the description in [PQ04],

$$\tau_e = \tau_{e0} + \tau_{e1}\exp\left(C_1\left(\frac{T_c}{300\,\text{K}} + C_0\right)^2 + C_2\left(\frac{T_c}{300\,\text{K}} + C_0\right) + C_3\frac{T_L}{300\,\text{K}}\right). \tag{A.9.0-12}$$

The hole energy relaxation time is set to 0.1 ps.

Table A.22: Energy relaxation time parameters

	$\mathrm{In_{1-x}Ga_xAs}$			$\mathrm{In_{1-x}Ga_xAs_yP_{1-y}},\, x = 0.47y$		
	a	b	c	a	b	c
$\tau_{e0}/(\mathrm{ps})$	1	0	0	1	0	0
$\tau_{e1}/(\mathrm{ps})$	100	0	0	100	0	0
C_0	3	-37	34	-18	37.3	-26.1
C_1	-0.53	0	0	-0.04	-0.13	0
C_2	0.853	0	0	0	0.853	0
C_3	0.5	0	0	0	0.5	0

APPENDIX B

Impact of relative permittivity on the collector junction capacitance

While the difference of the relative permittivity in compatible semiconductor materials is typically relatively small (c.f. tab. B.1), for the sake of completeness, this section presents how it can be taken into account in a multi-region capacitance model. For simplicity, voltage limiting is not considered in the equations and the shorthand ε_k is used for $\varepsilon_0\varepsilon_{rk}$.

Solving (3.4.2-14) while keeping in mind the displacement field continuity (cf. (3.4.2-11)) yields for the voltage across the junction

$$
\begin{aligned}
V_{\mathrm{D}n} - V_{\mathrm{j}} = {} & \left(\sum_{i=1}^{n} \frac{\mathrm{q}N_i}{\varepsilon_i} \frac{(x_i - x_{i-1})^2}{2} \right) + \\
& \left(\sum_{i=1}^{n-1} \frac{\mathrm{q}N_n}{\varepsilon_i} (x_i - x_{i-1})(x_n - x_{n-1}) \right) + \\
& \left(\sum_{i=1}^{n-2} \sum_{k=i+1}^{n-1} \frac{\mathrm{q}N_k}{\varepsilon_i} (x_i - x_{i-1})(x_k - x_{k-1}) \right),
\end{aligned}
\tag{B.0.0-1}
$$

which can be solved for x_n and inserted into (3.4.2-13). Then, the additional terms caused by the different doping and permittivity regions can be summarized in an effective built-in voltage in analogy to (3.4.2-17) as

Table B.1: Relative permittivities of common material systems.

Material 1	Material 2	$\varepsilon_{r,1}$	$\varepsilon_{r,2}$
InP	$InGa_{0.47}As_{0.53}$	12.5	13.9
Si	$Si_{0.7}Ge_{0.3}$	11.7	13.1
GaAs	$In_{0.49}Ga_{0.51}P$	12.9	11.8
GaAs	$Al_xGa_{1-x}As$	12.9	10.1...12.9

$$
V_{Dn,\text{eff}} = V_{Dn} - \frac{qN_n}{2\varepsilon_n} \left\{ \left[\left(\sum_{i=1}^{n} \frac{\varepsilon_n}{\varepsilon_i}(x_i - x_{i-1})^2 \right) - x_{n-1} \right]^2 + \right.
$$

$$
\left(\sum_{i=1}^{n-1} \frac{2\varepsilon_n}{\varepsilon_i}(x_i - x_{i-1})x_{n-1} - \frac{N_i\varepsilon}{N_n\varepsilon_i}(x_i - x_{i-1})^2 \right)
$$

$$
\left. \left(\sum_{i=1}^{n-2} \sum_{k=i+1}^{n-1} \frac{2N_k\varepsilon_n}{N_n\varepsilon_i}(x_i - x_{i-1})(x_k - x_{k-1}) \right) - x_{n-1}^2 \right\}. \tag{B.0.0-2}
$$

Thus, while the inclusion of ε_r differences adds significantly to the mathematical effort for the multi-region model, the result is ultimately identical, i.e. the introduction of effective built-in voltages for each region.

APPENDIX C

Supplementary figures for model verification

C.1 250 GHz InP process

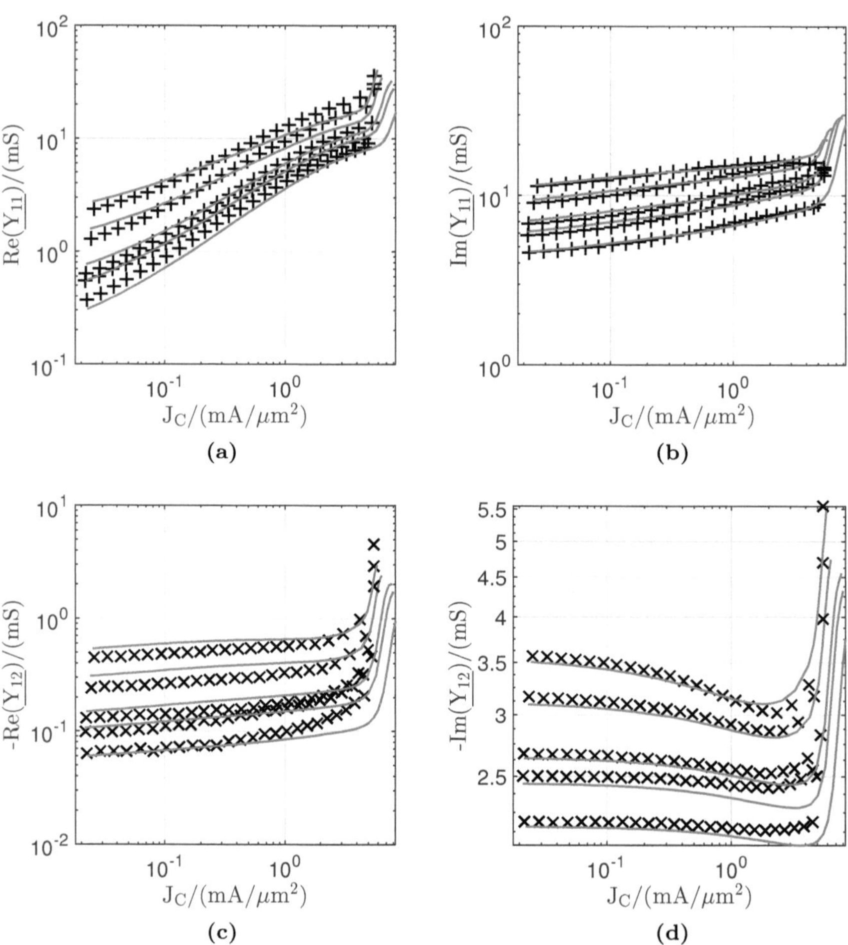

Figure C.1: Bias dependence of the Y-parameters of the devices with $l_{E0} = 15\,\mu m$, $b_{E0} = [0.5, 0.8, 1, 1.5, 2]\,\mu m$ at $V_{BC} = 0\,V$ and $f = 10\,GHz$. Lines: model and symbols: measurement. (a) $\mathrm{Re}\left(\underline{Y_{11}}\right)$, (b) $\mathrm{Im}\left(\underline{Y_{11}}\right)$, (c) $\mathrm{Re}\left(\underline{Y_{12}}\right)$, (d) $\mathrm{Im}\left(\underline{Y_{12}}\right)$.

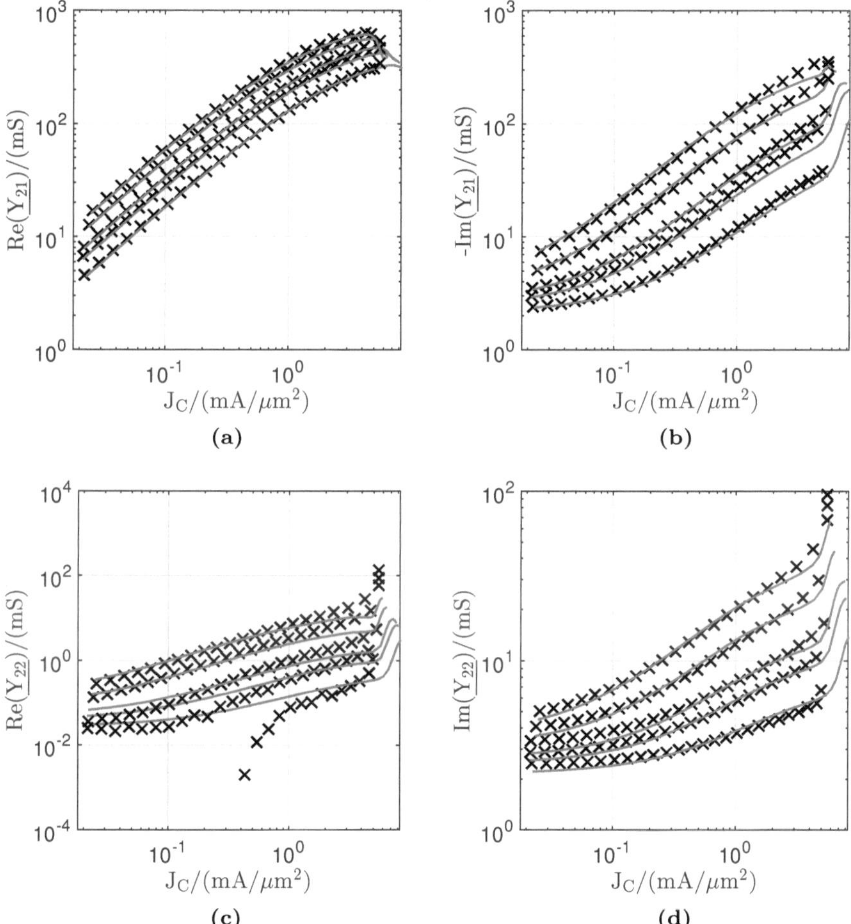

Figure C.2: Bias dependence of the Y-parameters of the devices with $l_{E0} = 15\,\mu m$, $b_{E0} = [0.5, 0.8, 1, 1.5, 2]\,\mu m$ at $V_{BC} = 0\,V$ and $f = 10\,GHz$. Lines: model and symbols: measurement. (a) Re $\left(\underline{Y_{21}}\right)$, (b) Im $\left(\underline{Y_{21}}\right)$, (c) Re $\left(\underline{Y_{22}}\right)$, (d) Im $\left(\underline{Y_{22}}\right)$.

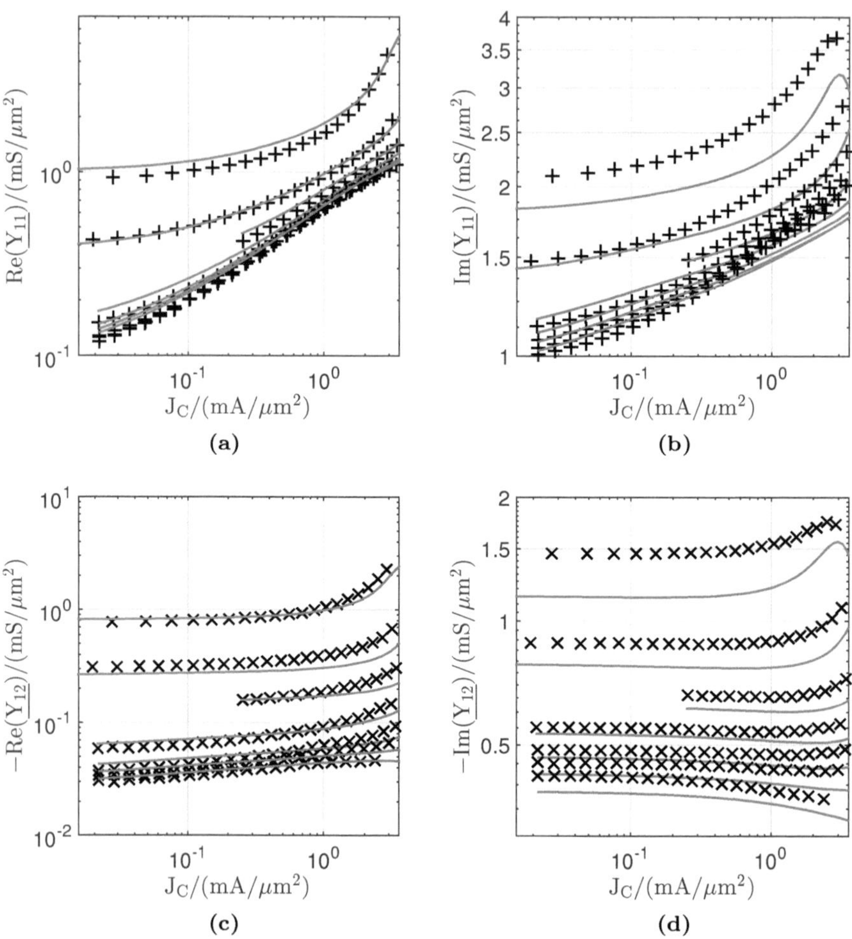

Figure C.3: Bias dependence of the Y-parameters of the reference device with $l_{\mathrm{E0}} = 15\,\mu\mathrm{m}$, $b_{\mathrm{E0}} = 0.8\,\mu\mathrm{m}$ at $V_{\mathrm{BC}} = [-0.25, 0, 0.1, 0.2, 0.3, 0.4, 0.5]\,\mathrm{V}$ and $f = 10\,\mathrm{GHz}$. Lines: model and symbols: measurement. (a) $\mathrm{Re}\left(\underline{Y_{11}}\right)$, (b) $\mathrm{Im}\left(\underline{Y_{11}}\right)$, (c) $\mathrm{Re}\left(\underline{Y_{12}}\right)$, (d) $\mathrm{Im}\left(\underline{Y_{12}}\right)$.

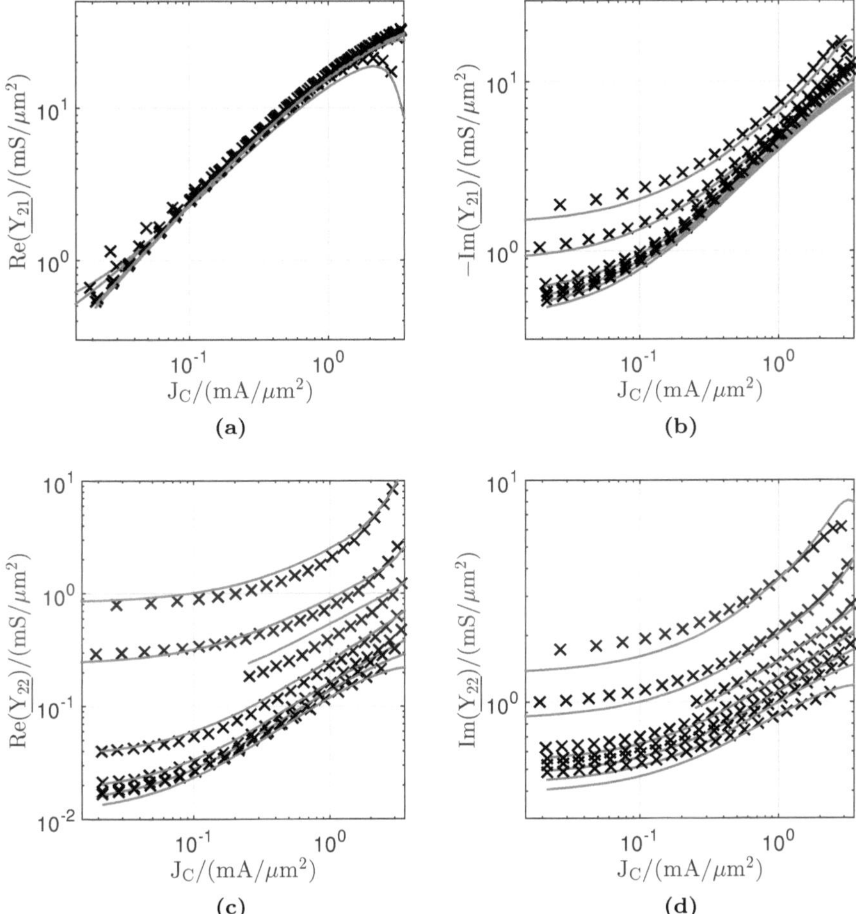

Figure C.4: Bias dependence of the Y-parameters of the reference device with $l_{E0} = 15\,\mu m$, $b_{E0} = 0.8\,\mu m$ at $V_{BC} = [-0.25, 0, 0.1, 0.2, 0.3, 0.4, 0.5]$ V and $f = 10\,\mathrm{GHz}$. Lines: model and symbols: measurement. (a) Re $(\underline{Y_{11}})$, (b) Im $(\underline{Y_{11}})$, (c) Re $(\underline{Y_{22}})$, (d) Im $(\underline{Y_{22}})$.

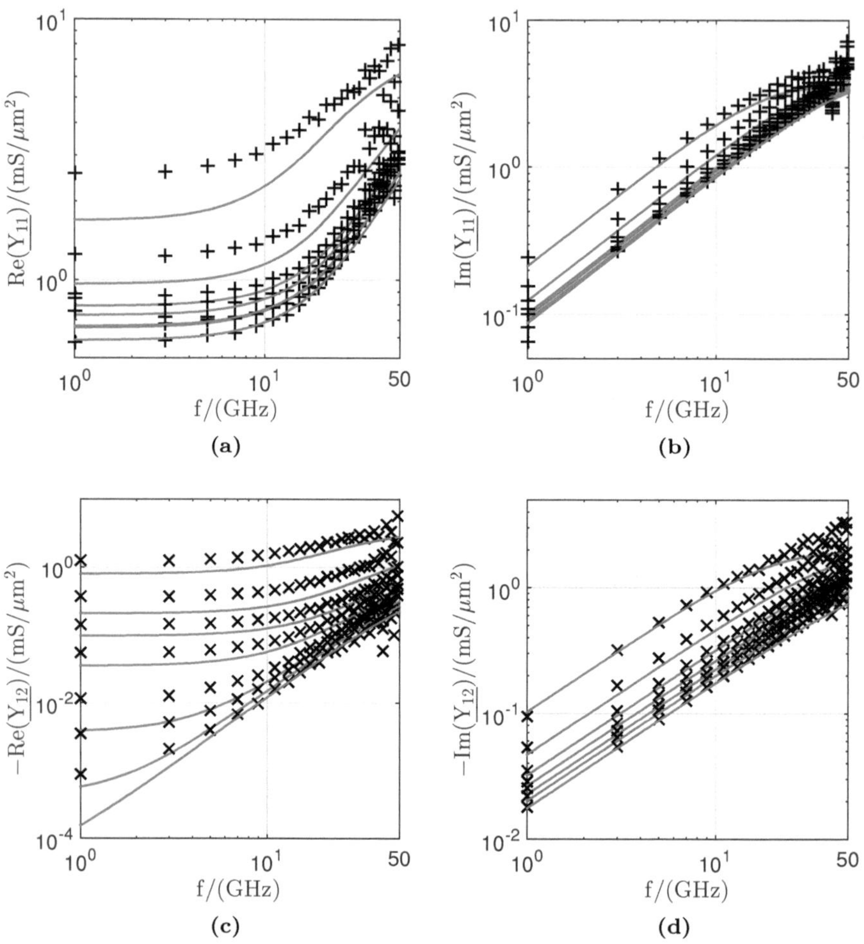

Figure C.5: Bias dependence of the Y-parameters of the reference device with $l_{E0} = 15\,\mu m$, $b_{E0} = 0.8\,\mu m$ at $V_{BC} = [-0.25, 0, 0.1, 0.2, 0.3, 0.4, 0.5]$ V and $J_C = 3\,mA\,\mu m^{-2}$ (highest recommended operating current). Lines: model and symbols: measurement. (a) $\mathrm{Re}\left(\underline{Y_{11}}\right)$, (b) $\mathrm{Im}\left(\underline{Y_{11}}\right)$, (c) $\mathrm{Re}\left(\underline{Y_{12}}\right)$, (d) $\mathrm{Im}\left(\underline{Y_{12}}\right)$.

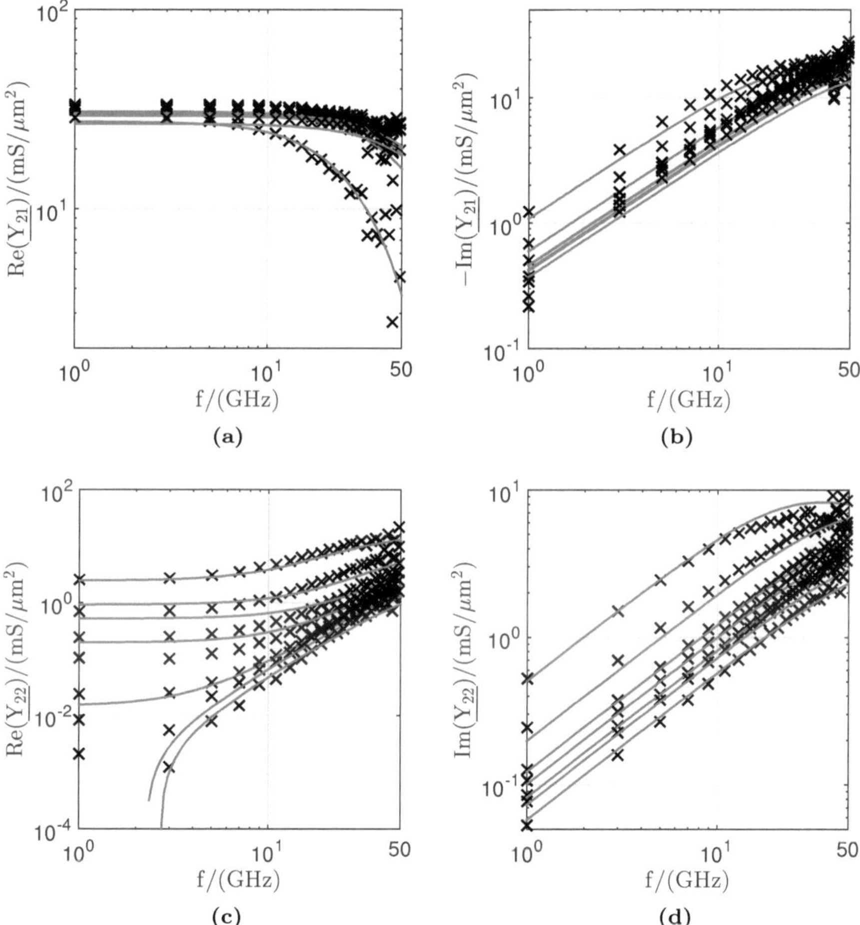

Figure C.6: Bias dependence of the Y-parameters of the reference device with $l_{E0} = 15\,\mu m$, $b_{E0} = 0.8\,\mu m$ at $V_{BC} = [-0.25, 0, 0.1, 0.2, 0.3, 0.4, 0.5]$ V and $J_C = 3\,mA\,\mu m^{-2}$ (highest recommended operating current). Lines: model and symbols: measurement. (a) $\mathrm{Re}\left(\underline{Y_{21}}\right)$, (b) $\mathrm{Im}\left(\underline{Y_{21}}\right)$, (c) $\mathrm{Re}\left(\underline{Y_{22}}\right)$, (d) $\mathrm{Im}\left(\underline{Y_{22}}\right)$.

C.2 500 GHz InP process

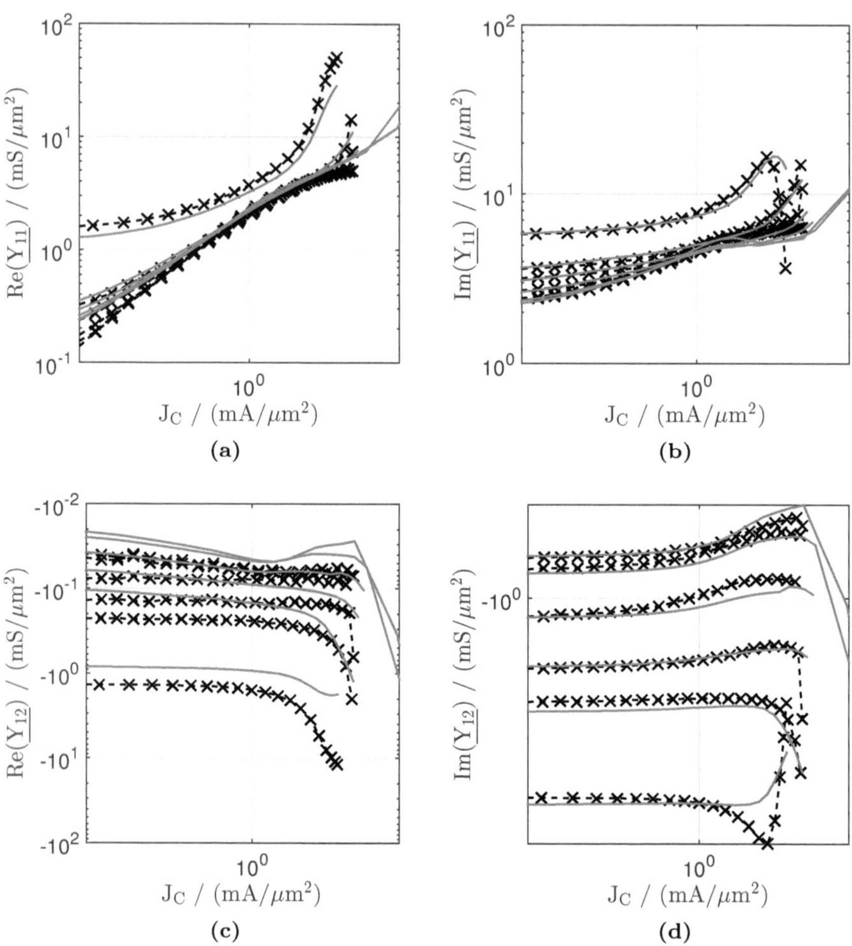

Figure C.7: Bias dependence of the Y-parameters of the reference device with $l_{E0} = 4\,\mu m$, $b_{E0} = 0.25\,\mu m$ at $V_{BC} = [-0.75, -0.5, -0.25, 0, 0.25, 0.5]$ V and $f = 10\,GHz$. Lines: model and symbols: measurement. (a) Re $(\underline{Y_{11}})$, (b) Im $(\underline{Y_{11}})$, (c) Re $(\underline{Y_{12}})$, (d) Im $(\underline{Y_{12}})$.

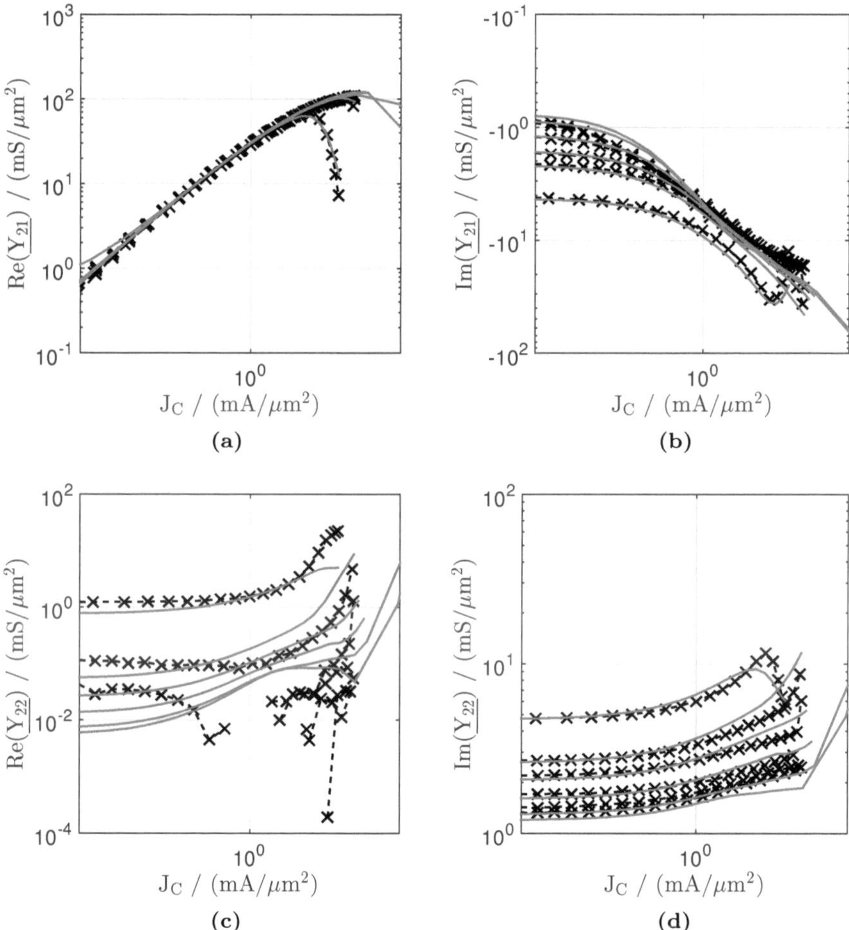

Figure C.8: Bias dependence of the Y-parameters of the reference device with $l_{E0} = 4\,\mu m$, $b_{E0} = 0.25\,\mu m$ at $V_{BC} = [-0.75, -0.5, -0.25, 0, 0.25, 0.5]$ V and $f = 10\,GHz$. Lines: model and symbols: measurement. (a) $\mathrm{Re}\left(\underline{Y_{21}}\right)$, (b) $\mathrm{Im}\left(\underline{Y_{21}}\right)$, (c) $\mathrm{Re}\left(\underline{Y_{22}}\right)$, (d) $\mathrm{Im}\left(\underline{Y_{22}}\right)$.

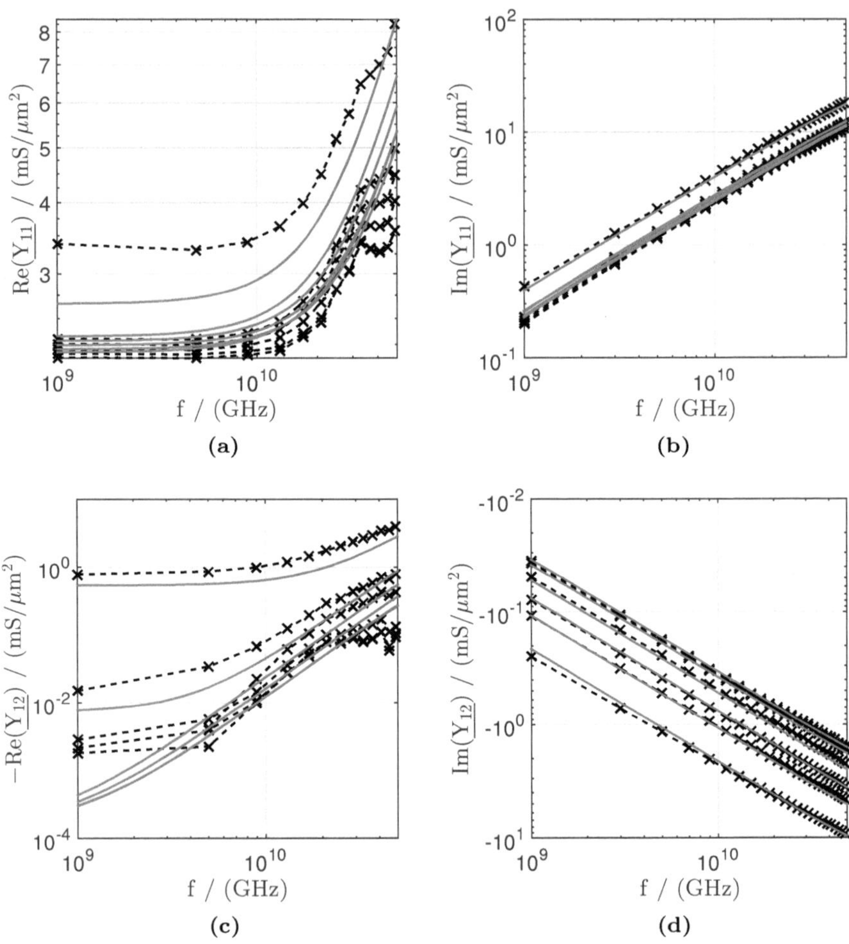

Figure C.9: Frequency dependence of the Y-parameters of the reference device with $l_{\mathrm{E0}} = 4\,\mu\mathrm{m}$, $b_{\mathrm{E0}} = 0.25\,\mu\mathrm{m}$ at $V_{\mathrm{BC}} = [-0.75, -0.5, -0.25, 0, 0.25, 0.5]$ V. Current density chosen so that transistor operates at $f_t/2$. Lines: model and symbols: measurement. (a) Re $\left(\underline{Y_{11}}\right)$, (b) Im $\left(\underline{Y_{11}}\right)$, (c) Re $\left(\underline{Y_{12}}\right)$, (d) Im $\left(\underline{Y_{12}}\right)$.

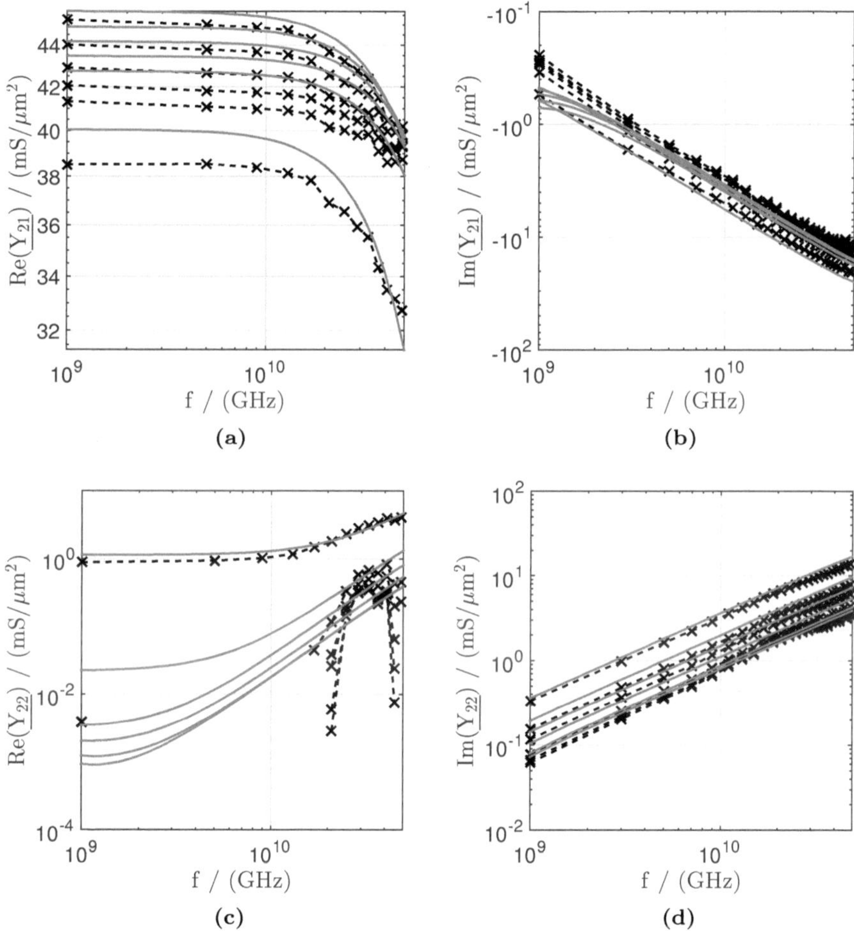

Figure C.10: Frequency dependence of the Y-parameters of the reference device with $l_{E0} = 4\,\mu\text{m}$, $b_{E0} = 0.25\,\mu\text{m}$ at $V_{BC} = [-0.75, -0.5, -0.25, 0, 0.25, 0.5]$ V. Current density chosen so that transistor operates at $f_t/2$. Lines: model and symbols: measurement. (a) $\text{Re}\left(\underline{Y_{21}}\right)$, (b) $\text{Im}\left(\underline{Y_{21}}\right)$, (c) $\text{Re}\left(\underline{Y_{22}}\right)$, (d) $\text{Im}\left(\underline{Y_{22}}\right)$.

C.3 GaAs process

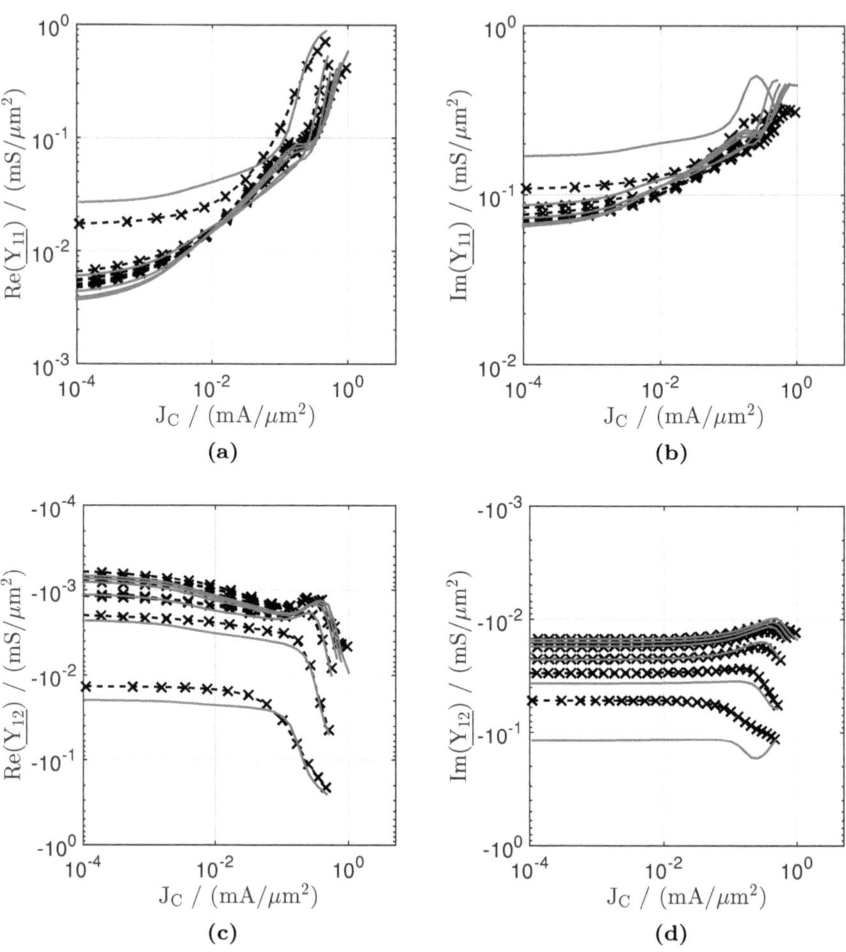

Figure C.11: Bias dependence of the Y-parameters of the GaAs device at $V_{\mathrm{BC}} = [-0.5, -0.25, 0, 0.5, 0.8, 1.0]$ V and $f = 3\,\mathrm{GHz}$. Lines: model and symbols: measurement. (a) Re $(\underline{Y_{11}})$, (b) Im $(\underline{Y_{11}})$, (c) Re $(\underline{Y_{12}})$, (d) Im $(\underline{Y_{12}})$.

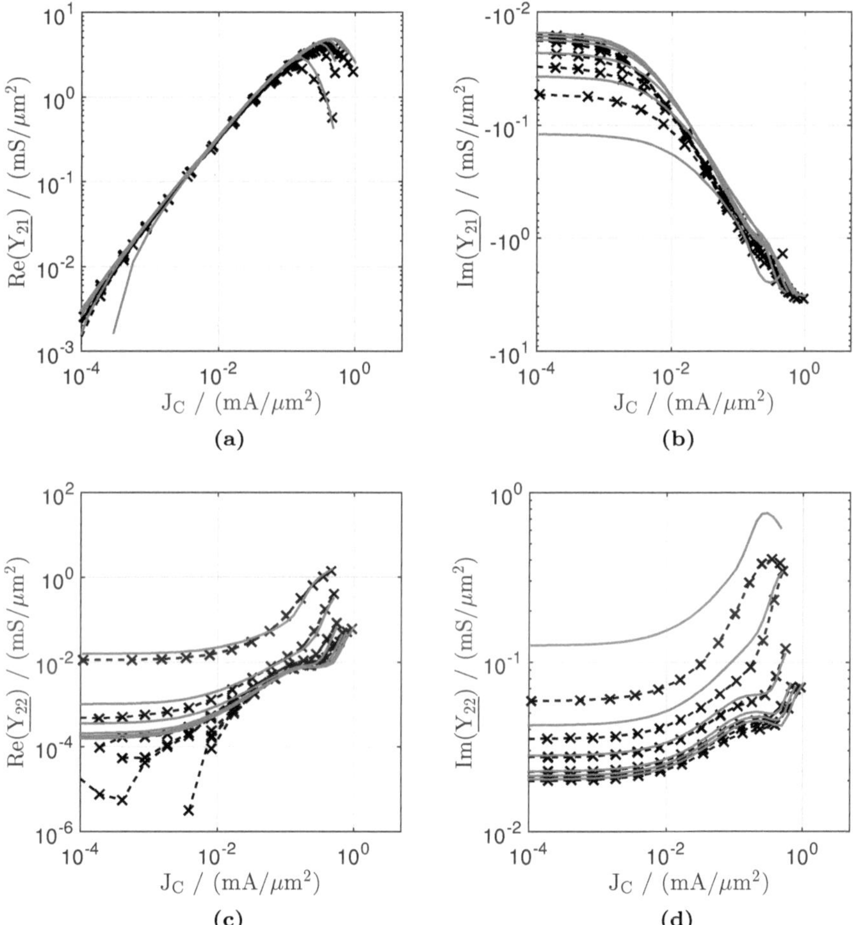

Figure C.12: Bias dependence of the Y-parameters of the GaAs device at $V_{\mathrm{BC}} = [-0.5, -0.25, 0, 0.5, 0.8, 1.0]$ V and $f = 3\,\mathrm{GHz}$. Lines: model and symbols: measurement. (a) $\mathrm{Re}\left(\underline{Y_{21}}\right)$, (b) $\mathrm{Im}\left(\underline{Y_{21}}\right)$, (c) $\mathrm{Re}\left(\underline{Y_{22}}\right)$, (d) $\mathrm{Im}\left(\underline{Y_{22}}\right)$.

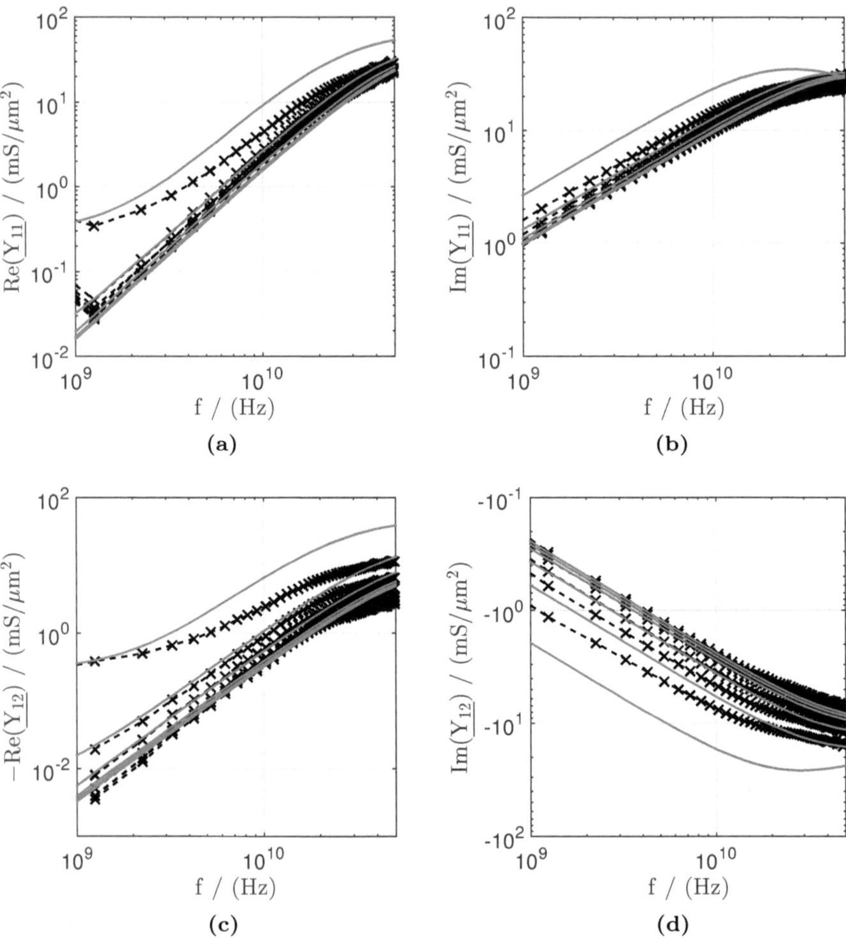

Figure C.13: Frequency dependence of the Y-parameters of the GaAs device at $V_{\mathrm{BC}} = [-0.5, -0.25, 0, 0.5, 0.8, 1.0]$ V. Current density chosen so that transistor operates at $f_t/2$. Lines: model and symbols: measurement. (a) Re $\left(\underline{Y_{11}}\right)$, (b) Im $\left(\underline{Y_{11}}\right)$, (c) Re $\left(\underline{Y_{12}}\right)$, (d) Im $\left(\underline{Y_{12}}\right)$.

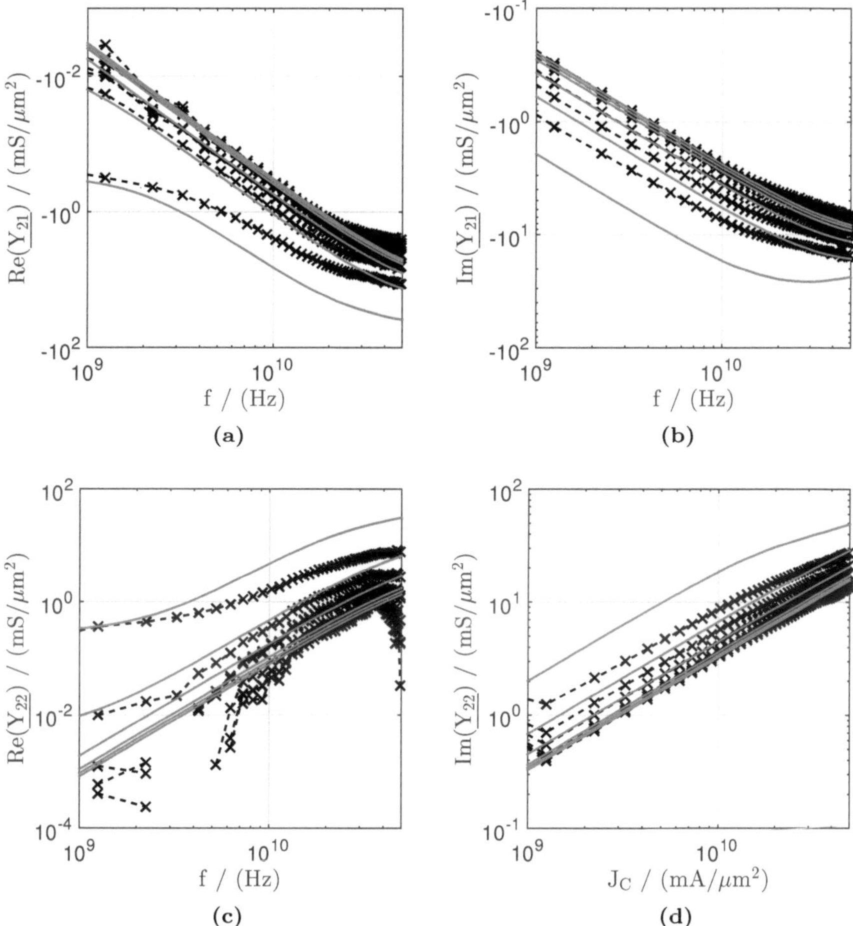

Figure C.14: Frequency dependence of the Y-parameters of the GaAs device at $V_{\mathrm{BC}} = [-0.5, -0.25, 0, 0.5, 0.8, 1.0]$ V. Current density chosen so that transistor operates at $f_t/2$. Lines: model and symbols: measurement. (a) Re $\left(\underline{Y_{21}}\right)$, (b) Im $\left(\underline{Y_{21}}\right)$, (c) Re $\left(\underline{Y_{22}}\right)$, (d) Im $\left(\underline{Y_{22}}\right)$.